Cloud Technologies

Cloud Technologies

An Overview of Cloud Computing Technologies for Managers

Roger McHaney
Kansas State University
Manhattan
USA

Registered Offices
John Wiley & Sons, Inc., 111 River Street, Hoboken, NJ 07030, USA
John Wiley & Sons Ltd, The Atrium, Southern Gate, Chichester, West Sussex, PO19 8SQ, UK

Editorial Office
The Atrium, Southern Gate, Chichester, West Sussex, PO19 8SQ, UK

For details of our global editorial offices, customer services, and more information about Wiley products visit us at www.wiley.com.

Wiley also publishes its books in a variety of electronic formats and by print-on-demand. Some content that appears in standard print versions of this book may not be available in other formats.

Library of Congress Cataloging-in-Publication Data

Names: McHaney, Roger, author.
Title: Cloud technologies : an overview of cloud computing technologies for
    managers / Roger McHaney, Kansas State University, Manhattan, USA.
Description: First edition. | Hoboken, NJ : Wiley, 2021. | Includes
    bibliographical references and index.
Identifiers: LCCN 2020040717 (print) | LCCN 2020040718 (ebook) | ISBN
    9781119769521 (cloth) | ISBN 9781119769491 (adobe pdf) | ISBN
    9781119769507 (epub)
Subjects: LCSH: Cloud computing. | Organizational change. | Information
    technology–Management. | Business enterprises–Computer networks.
Classification: LCC QA76.585 .M425 2021 (print) | LCC QA76.585 (ebook) |
    DDC 004.67/82–dc23
LC record available at https://lccn.loc.gov/2020040717
LC ebook record available at https://lccn.loc.gov/2020040718

Cover Design: Wiley
Cover Image: © zf L/Getty Images

Set in 9.5/12.5pt STIXTwoText by SPi Global, Chennai, India
Printed and bound by CPI Group (UK) Ltd, Croydon, CR0 4YY

C9781119769521_160321

Contents

Preface

Cloud computing resulted from the perfect storm (to continue the weather paradigm). With growing numbers of software systems, databases, security requirements, hardware needs, and so forth, IT specialists had trouble innovating while simultaneously managing daily operations. Being an expert in so many areas was daunting and beyond the capabilities of many Information Technology (IT) departments. To make things even more difficult, technologies changed continually. Hardware and software updates needed to be rolled out across organizations that were spread out geographically. Chief Information Officers (CIOs) and IT managers looked for ways to reduce the burden and provide services to end users who needed to do their jobs.

At the same time, the Internet matured to the point where high-speed connections meant increasingly complex applications worked remotely. This opened the door to many new possibilities. Experts could be located anywhere that had an Internet connection. Suddenly, cloud computing concepts enabled IT specialists to offer new technologies within their organization without needing an onsite expert with each implementation. The cloud provided access to custom tool sets, even if the tools were used only on rare occasion. This meant IT departments and users could cut costs and return their focus to core business needs. Instead of IT driving the business, the business used IT to remove technology-related obstacles and ultimately achieve success with its primary mission.

Cloud computing changes how organizational planners think about and implement IT infrastructure. Any manager without a firm grasp of the cloud's fundamental concepts and potential is at a huge disadvantage in the modern world. *Cloud Technologies: An Overview of Cloud Computing Technologies for Managers* describes cloud computing concepts and ways these enhance strategies and operations in language free of technobabble. Topics important to managers, such as cost savings and enhanced capabilities, together with different models for implementing cloud technologies, are described. Specifically, *Cloud Technologies* introduces cloud-based systems and how these technologies change the ways organizations approach and implement their computing infrastructure. This book starts with a high-level overview of cloud computing then moves into business related considerations such as Service Level Agreements, elasticity, security, audits, and other practical implementation issues. Additional topics such as automation, infrastructure as code, DevOps, Fog computing, and orchestration also are covered.

I hope you enjoy reading this book. I learned a great deal while writing it and hope you benefit from its content. The most successful readers will use this as a starting point and continue their journey as new cloud computing technologies emerge and the landscape continues to evolve.

Ever Onward, Professor Roger McHaney

Acknowledgments

Teachers learn from their students and I want to acknowledge the incredible contributions made by my BIPOC students. The insights, patience, wisdom, and viewpoints they have shared with me over the years influence my thinking in ways they cannot imagine. THANK YOU!

The Kansas State University College of Business provided funding for my sabbatical which enabled the development of this book. Thank you!

The manuscript was developed in Edinburgh, Scotland. The city, its inspirations, and resources greatly influenced and enhanced this book project. Thanks to Dickins Edinburgh Ltd. for finding my wife and I the perfect home during our time there.

Finally, a special thanks to my wife, Annette, for keeping me connected, grounded, and being part of the adventure that resulted in this book.

Ever Onward, Roger McHaney

About the Companion Website

This book is accompanied by a companion website:

www.wiley.com/go/mchaney/cloudtechnologies

The website includes PowerPoint slides.

1

What Is Cloud Computing?

Cloud computing is a broad term that simply means delivering computing services – which may include servers, databases, storage, networking, software, data analytics, security solutions, organizational systems, virtual computers, and much more – over the Internet. The term "cloud" came from a symbol in old flow charts and diagrams used to represent the Internet (see Figure 1.1). This symbol suggested resources exist, but we may not necessarily know exactly where they are or how they work. But, through Internet connections and Web-based interfaces, we can set up and configure exactly what we need to use. So, cloud computing becomes the mindset for shared resources that can be used economically from just about anywhere.

> The phrase "Cloud Computing" originates from the "Cloud" symbol used in flow charts and diagrams to represent the Internet. The idea is that any computer with a Web interface has access to an incredible pool of computing resources, power, applications, and files.

Why Cloud Computing?

Cloud computing was the result of the perfect storm to continue the atmospheric paradigm. With the growing number of software systems, databases, security requirements, hardware needs, and so forth, IT specialists found it difficult to keep up. Becoming an expert in so many areas was daunting and beyond the capabilities of many Information Technology (IT) departments. Technologies changed continually. Hardware and software updates needed to be rolled out across organizations that often were spread out geographically. Chief Information Officers (CIOs) and IT managers looked for ways to reduce the burden and provide high end services to end users who needed to do their jobs.

At the same time, the Internet was maturing to the point where high-speed connections meant increasingly complex applications worked remotely. This opened the door to many new possibilities. Experts could be located anywhere that had an Internet connection. Suddenly, cloud computing concepts enabled IT specialists to offer new technologies within their organization without needing in-depth knowledge about, or high levels of expertise with each implementation. The cloud provided access to custom tool sets, even if the tools

Cloud Technologies: An Overview of Cloud Computing Technologies for Managers, First Edition. Roger McHaney.
© 2021 John Wiley & Sons Ltd. Published 2021 by John Wiley & Sons Ltd.
Companion website: www.wiley.com/go/mchaney/cloudtechnologies

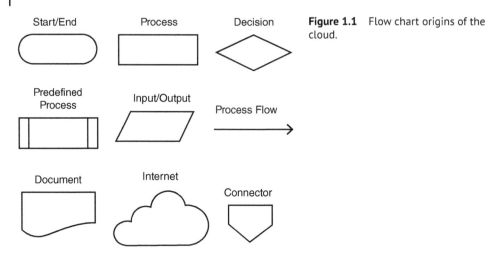

Figure 1.1 Flow chart origins of the cloud.

were used only on rare occasion. This allowed IT departments and users to cut costs and returned the focus to core business needs. Instead of IT driving the business, the business used IT to achieve success with its primary mission and to remove technology-related obstacles.

Cloud Computing's Focus

The benefits of cloud computing helped this new concept win rapid acceptance among Senior IT Managers (although lingering concerns remain which we will cover later). Today, cloud computing focuses on three primary areas: *cost reduction*, *capacity planning*, and *business agility*. All three areas relate to business needs and help make a case for moving to the cloud (see Figure 1.2).

Cost Reduction

IT costs generally come in two categories: (i) cost of new equipment/software (e.g. purchase costs); and (ii) ongoing costs of ownership (e.g. operational costs). As equipment ages and maintenance becomes more time consuming, operational costs can quickly escalate.

In many organizations, it is important to directly tie the costs of IT items to their related business uses. In other words, managers want to know how much it costs to accomplish business outcomes so the profitability of certain activities can be monitored. This alignment between IT costs and business performance can be very difficult to both understand and maintain. Several factors come into play here. First, software and hardware may be used in many ways and for many purposes across an organization. Second, IT expenditures often relate to what internal users expect their maximum usage to eventually become.

Think of it in this way. You have just graduated with your university degree. Your smart boy or girl friend sees the future value of your education and decides to make the relationship more permanent by proposing marriage (they love you as well I am sure!). You think ahead a little and decide to invest in your dream home. Even though it is just the two

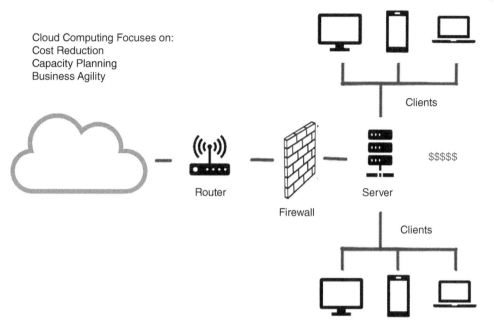

Cloud Computing Focuses on:
Cost Reduction
Capacity Planning
Business Agility

Clients

Router

Firewall

Server

$$$$$

Clients

Figure 1.2 The focus of cloud computing.

of you and you could happily live in a single room flat, you decide to buy a four bedroom mini-mansion with a swimming pool. The idea is that in the future, you will need bedrooms for your 2.5 children and a place for your mother-in-law to stay when she visits. If you were trying to allocate the costs, it would be a bit of a challenge. You do not have children yet, but you "bought" space where they will eventually sleep.

Organizations face the same challenge. Do you buy for the future to leverage current purchasing power? Or do you wait until the last minute and face challenges that may include higher costs, slow implementation, and lack of availability? So, you can see the dilemma that many organizations face.

Figure 1.3 illustrates how purchases of software and hardware comprise part of a firm's total IT costs, but the ongoing costs can climb, particularly when a new item is acquired (represented by the stair step in "Purchases"). As items become older, often the cost of maintenance increases, sometimes in ways that were unexpected.

In most cases, operational overhead accounts for a large percent of IT budgets. Over time, these costs probably will exceed up-front investment and purchase costs. Think about everything that goes into operating expenses:

- Technical personnel with high levels of skills. This means regular training, certification and other costs are required.
- Software and hardware upgrades and patches that need testing, user training, and installation time.
- Software leasing costs. Many software packages require a yearly fee to pay for upgrades and so forth.

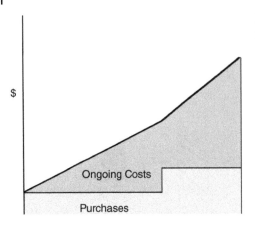

Figure 1.3 Ongoing costs and purchases as part of total cost.

- Operation costs like electric utility bills and investments for cooling systems, surge protection, air filtration systems and so forth.
- The costs of security and access control measures must be in place. These operations are extremely important and include software, network, and physical infrastructure components (e.g. virus protection, intrusion detection, video cameras, locks, and so forth).
- Help desk, staff specialists, and administrative assistants might be required to track licensing, do training and provide daily assistance.
- In addition, many other expenses also exist (like insurance, audits, travel, staff meetings and retreats, et cetera).

The costs of IT, since these are difficult to tie directly to a business activity, may be viewed as a cost sink. This often places IT expenditures right in the crosshairs of managers responsible for cutting costs and reducing budgets. So, IT managers must both innovate and cut costs. That can become a bit of a nightmare. Luckily, cloud computing can help eliminate many of these problems and we will investigate these advantages in Chapter 6.

Capacity Planning

Another area driving IT leaders into the cloud relates to capacity planning. This is the process of deciding how to prepare for the future. In the example of the house purchase, this would include discussions about how many children the happy couple might like to eventually have, how often visitors are expected, and other aspects of how future life is expected to unfold. As you can imagine, often planning and dreaming ends up a long way from reality.

In the IT world, similar activities occur. Capacity planning is the process used to determine and fulfill future demands expected to be placed on an organization's IT resources, duties, operations, and services. Often, the CIO or another senior IT leader is part of an organization's strategic planning group and works to understand the maximum and type of IT demand that may emerge. This figure is used to determine required resources and if existing IT assets are capable of meeting requirements for a specific time. Differences between perceived needs and existing capacity can have several results. If a discrepancy exists, a system might become overwhelmed, also called under-provisioned. Under-provisioned systems cannot meet the needs of the users. If the opposite situation exists and too much IT capacity exists, the system is considered over-provisioned. This is inefficiency and causes waste. Having too much capacity results in waste and money that could have been spent in other, more productive ways. Capacity planning focuses on minimizing any discrepancy, whether it results in over- or under-provisioning and seeks to achieve predictable efficiency and optimal performance levels.

Capacity planning can be approached from a variety of perspectives. Three popular approaches to capacity planning strategy are:

- *Match Strategy* which involves adding IT resources in small increments to match demand as best as possible. This strategy avoids over-provisioning but sometimes fails to take advantage of volume discounts. It also can result in systems that are not able to ramp up quickly enough (e.g. adding staff, training, and so forth).
- *Lead Strategy* means that IT resource capacity is added in anticipation of demand. This strategy can take advantage of volume purchases but is more susceptible to over-provisioning and can result in unused resources that become obsolete before they are needed.
- *Lag Strategy* adds IT resources only after capacity is reached. This approach can suffer from under-provisioning and from being unable to rapidly react to organizational needs. It may be best in situations where demand for IT resources is steady and less susceptible to unexpected changes.

Capacity planning often is considered one of a CIO's most difficult challenges. Finding the sweet spot between too much and too little is difficult and requires understanding the business and its strategic plans and then matching that with knowledge of the IT world, together with a sense of where major systems and capabilities are heading in the future. Many CIOs have lost their jobs because they anticipated things that did not happen or were swept away by broader technology changes. Attending conferences and trade shows can help an IT planner avoid making big mistakes but foretelling the future is never easy. Despite these challenges, a bright spot exists: the cloud makes capacity planning much easier, as we will see in this text.

Organizational Agility

Another major business driver making the cloud approach to computing attractive is organizational agility. The cloud allows businesses to easily acquire and roll out the latest technologies without needing to develop as much internal expertise. More than ever, successful businesses need the ability to change rapidly to innovate and compete. External technology evolves, competitive forces emerge, and internal needs force change on an organization. Agility is the measure of how responsive an organization can be when it comes to these changes.

A big part of organizational agility is called elasticity. IT departments must be ready to respond to business changes by upsizing or downsizing capabilities to match the current situation. This is a little different than capacity planning because it involves a time element. How fast can an IT system resize to meet demand? And, often this is in response to changes in scope that are much different than what was expected or planned.

If we go back to our happy family example, think about the instance where following a visit to the doctor, the birth of an expected daughter or son turned out to be triplets. Agility would be how fast preparations can be made. Do more cribs need to be purchased? Do babysitting services need to be rethought? Is a new house needed? So, capacity planning is ongoing and necessary, and it must be done in a timely fashion (a few months for our example).

Businesses are no different. Perhaps a new product goes viral and demand outstrips production. Or a company website becomes incredibly popular and its servers need to be upgraded to ensure people are not frustrated when they visit. If they are purchasing a product on the site, this becomes even more crucial. The IT department may have little time to respond because viral popularity could come and go fast. And then later, maybe things need to be ramped down again to prevent being over-provisioned which preserves capital and reduces waste.

In many cases, particularly when non-cloud, in-house solutions are used, capacity changes can be difficult, or perhaps not possible. Cloud-based solutions offer businesses more flexibility when it comes to responding to change because of their inherent scalability. If a business does not have the ability to scale its solutions rapidly and tries anyway, the result could be disastrous. For instance, reliability might suffer, customers could be lost and opportunities for new competitors might emerge.

> The term "Scalability" is associated with computing solutions. When a system is scalable, it can grow without being hampered by existing structure or available resources. A scalable system can respond to higher demand with little or no changes required.

How Is Cloud Computing Hosted?

Cloud computing has become the standard way for companies to access IT infrastructure, hardware, and software resources. A cloud can be hosted specifically by an organization for its exclusive use, or an organization can use services that reside on a cloud hosted by an outside vendor. These two deployment approaches, called private and public clouds, each have pros and cons. Some organizations use a third approach called a hybrid cloud.

Private Cloud Deployment

A private cloud, also called an internal or enterprise cloud, is hosted on hardware systems located within an organization's data center. A private cloud is essentially an intranet where all the data and services are protected by a firewall. Private clouds are generally used by large organizations that may already have data centers or extensive IT expertise. The benefit is that all data is held by the organization on its premises and can offer customized security solutions. This might be preferable to organizations that manage critical and confidential data. The biggest drawback is that an expert, internal IT group is responsible for all data management, software updates, hardware maintenance, and so forth.

Public Cloud Deployment

An organization that opts to use a public cloud has turned the day-to-day management of hosting over to a third-party vendor. Generally, the organization using the public cloud is

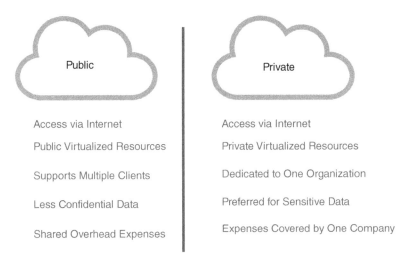

Figure 1.4 Public versus private clouds.

not responsible for managing the system, updating software, and maintaining the hardware. Instead, data and applications are stored in the host's data center where all IT operations are managed. Using a public cloud can reduce IT overhead because these functions are taken care of by someone else with expertise in the area, and that shares the costs among several client organizations. Some organizations avoid public clouds because their data is no longer on their own premises and they worry that security problems could result. While this is possible, it is not any more likely than data security breaches on private clouds. Most public cloud hosts have excellent security and maintain separate areas for each of their clients' data and applications (see Figure 1.4).

Hybrid Cloud Deployment

Other organizations need features offered by both public and private clouds. Perhaps their sensitive data must be held on-premise or they use a custom application that requires internal hosting. A hybrid cloud combines features of both public and private clouds, using technology that permits sharing data and applications between them. A hybrid cloud gives a business greater flexibility and the ability to deploy applications and data in more ways.

What Are the Different Types of Cloud Solutions?

Whether a cloud is private or public, it can be used for a variety of functions. It can replace hardware, software, organizational computing infrastructure, and many other IT resources. Cloud computing can be simple. For example, using a service like Dropbox for file sharing. Or it can be complex. For example, a multinational Fortune 500 company using an enterprise resource planning system like SAP to run its operations in 16 different countries. Despite the wide range of possibilities, most solutions fit into one of three general categories. These are Software as a Service (SaaS), Platform as a Service (PaaS),

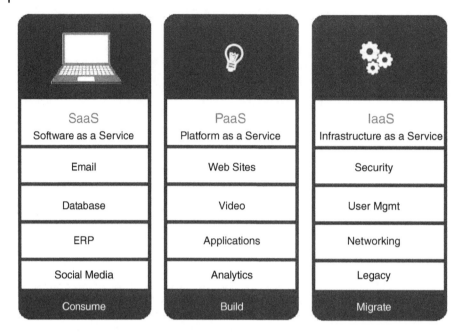

SaaS	PaaS	IaaS
Software as a Service	Platform as a Service	Infrastructure as a Service
Email	Web Sites	Security
Database	Video	User Mgmt
ERP	Applications	Networking
Social Media	Analytics	Legacy
Consume	Build	Migrate

Figure 1.5   Three general cloud solution areas.

and Infrastructure as a Service (IaaS). The type of cloud solution reflects the complexity of the systems and how many functions it can replace. The next sections describe these three general cloud solutions in more detail. Figure 1.5 illustrates our starting point.

Software as a Service (SaaS)

SaaS usually first comes to mind when people talk about cloud computing. It started as an Internet-based software distribution solution. Not too long ago, software was installed from disks, CDs or DVDs directly onto user desktops. As the Internet became more robust and faster, SaaS was the natural next step. Software ran on host computers accessed via the Internet. This ensures customers have the latest edition of software and more easily manage their payment and use options. SaaS is sometimes called on-demand or hosted software.

In the SaaS approach, a software vendor hosts the application at their data center and, in most cases, a customer accesses it via a Web browser or a custom interface that has been downloaded to the user's computer. SaaS applications generally target business users, although more and more we see companies going directly to private consumers. In the business world, examples of SaaS include subscription-based software like Salesforce or SAS; and pay-per-use software like WebEx which charges for online meetings. Other services, like Dropbox charge a yearly fee that is prorated, based on level of service.

SaaS is currently the most popular software distribution model for many business applications, including document management, accounting, human resource management, help desk management, content management, collaboration, and other core business areas. Thousands of SaaS vendors offer 10's of thousands of business applications, made possible by the Internet. Most SaaS vendors use the following approaches to manage their software:

- Software updates applied automatically and transparently to the end-user.
- Users pay a subscription fee based on number of seats, level of service, storage requirements or other criteria.
- The only hardware required by the customer is a desktop, laptop, or mobile device capable of connecting over the Internet. In most cases, no extra hardware is sent to the customer.

Benefits of SaaS Solutions

- Elasticity: Customers can quickly and seamlessly add users, capacity, or capabilities. Often, this is automatic.
- Accessibility: End-users can access the software from anywhere that an Internet connection exists. This is particularly helpful for those who travel or are on the road for their work.
- Cost Savings: Costly infrastructure and in-house expertise costs can be minimized or even eliminated. These are built into subscription fees.

Platform as a Service (PaaS)

PaaS is related to SaaS but is a broader and more complex form of cloud computing generally used by organizations involved in custom application development. Usually, PaaS solutions make it easier to develop, test, collaborate, maintain, track releases and changes, and perform other duties related to software creation. PaaS generally offers a configured sandbox for software testing, project management features, and a deployment environment that enables customers to roll out their cloud-based applications. Examples of commercial PaaS clouds include Amazon Web Services (AWS) and Microsoft Azure.

PaaS usually includes a solution or software stack for development. The term solution stack refers to a set of software components that are required to complete an entire application. This means no further components are needed to create a complete platform. Applications run on top of the resulting platform. Consider a web application as an example. The developer would need to define the stack as the operating system, web server, database, and programming/development language. In the case of business applications, a stack might consist of the operating system, middleware, database, and end-user application. In a development environment, components within the stack often are created by different experts independently.

Commonly used stacks are given acronyms that represent the components. Figure 1.6 provides an example of the WINS stack. WINS comprises Windows Server for the operating system, Internet Information Services as the Web servers, .NET as the framework for software development, and SQL Server to run the required database.

The terms *software stack*, *solution stack* and just *stack* often are used interchangeably by developers and IT specialists. However, the term solution stack may include hardware components in the overall solution. In cloud-based environments, these have become virtualized and integrated into the PaaS cloud solution.

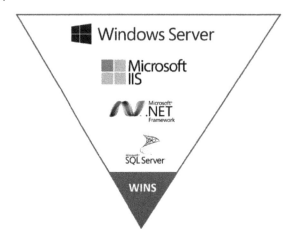

Figure 1.6 WINS solution stack.

Other Example Stacks

WIMP: Windows (operating system); Internet Information Services (web server); MySQL (database); PHP, Perl, or Python (programming language)

WISA: Windows Server (operating system); Internet Information Services (web server); SQL Server (database); ASP.NET (web framework)

PaaS clouds offer several beneficial advantages to organizations. Foremost is the sense of a community formed by an organization's developers. Most modern applications require the interaction and cooperation of many specialists. The idea that an online community is facilitated within the PaaS cloud is an excellent benefit of the technology. Another plus in PaaS, which is common to most cloud services, is that an organization is not required to update its infrastructure software. The PaaS vendor manages all updates, fixes, patches, and regular software maintenance activities. This reduces operational costs. Likewise, upfront investment costs are reduced. Finally, the software development team spends its time building applications rather than maintaining, updating, and working on routine tasks associated with the testing and deployment environment.

Infrastructure as a Service (IaaS)

IaaS can be viewed as the lowest level of cloud solution and focuses more on system configuration issues. Specifically, IaaS clouds host infrastructure components traditionally managed in on-premise data centers. IaaS offers services that include servers, data storage, networking hardware, and virtual machines. IaaS seeks to be a fully outsourced service that replaces a data center. IaaS offers pre-configured hardware (or software) through an interface. Customers install software and services on the IaaS cloud and run/manage their applications as if it were an on-premise data center.

An IaaS model adds value by providing services that accompany infrastructure components. Often, these include billing, activity/load monitoring, access logs, security, load balancing, backup and recovery, replication, and other safety measures. IaaS services seek to be policy-driven, which means customers can automate many operations. Customers access services and are allocated resources using an Internet interface in most cases, although for more secure operations, a wide area network (WAN) or virtual tunnel may be used. An IaaS user may log into the cloud to install their own applications or software stack components or configure virtual machines (VMs), deploy middleware, create backups, and install enterprise applications. Customers also rely on the IaaS provider to monitor application and server performance, track costs, balance network traffic, manage disaster recovery, and monitor security.

> The term **middleware** is used to describe the software layer between the operating system and applications on each side of a distributed computing system in a network. Common middleware applications include web servers, application servers, content management systems, databases, and tools, such as ODBC, that enable database integrations.

Examples of organizations that provide IaaS services are Google (with its Google Cloud Platform), RackSpace, and Amazon with AWS. Many cloud providers offer both PaaS and IaaS services, AWS is one such example.

IaaS can be less expensive, faster, and more cost-efficient than developing an in-house data center. IaaS allows businesses to rent or lease infrastructure to avoid over- or under-provisioning their operations. IaaS is particularly useful for organizations that see fluctuations in their workload or projects. For companies that offer software services, being able to expand their infrastructure when a new project is under development and then release that infrastructure when complete, makes more sense than purchasing hardware that soon will become obsolete.

IaaS flexibility is reflected in the payment models used by most vendors. Generally, IaaS clients are charged on a usage basis. Pay-as-you-go models are currently the most common approach with billing based on transactions, time, or virtual space used. For customers deploying in an IaaS environment for the first time, billing can become an issue. Sometimes, the rates sound inexpensive but add up quickly. Since every activity on the system may incur a cost, the overall expense might be more than expected. It is also important for users to ensure their billing is accurate and no unexpected or unwanted services are running on the site.

Benefits of IaaS Solutions

- Reduces capital expenditures and outlays
- Can reduce overall cost of IT function
- Users only pay for the services needed
- Enterprise-grade IT resources and infrastructure are available even to small organizations
- Scalability and elasticity are very easy
- Users maintain control over their own application deployment if critical to their business model

SaaS versus PaaS versus IaaS: A Review

In general, cloud computing allows users to share overhead resources to achieve an economy of scale. In some ways, this is like a utility company which supplies electricity to a variety of customers. Everyone pays a portion of the upkeep expenses while experts ensure electricity is available and so forth.

All forms of cloud computing deliver services – software, databases, storage, servers, networking, and more over the Internet to users' organizations. The three primary forms of services are:

- **Infrastructure as a Service (IaaS):** *Hardware is supplied and managed by an external party.*
- **Platform as a Service (PaaS):** *Hardware and operating system are supplied and managed by an external party. The hardware functions are transparent to the user.*
- **Software as a Service (SaaS):** *Hardware, operating system, and applications are supplied and managed by an external party. Everything but the applications are transparent to the user.*

In a way, PaaS builds on IaaS because in addition to the hardware components, it provides and manages operating systems, middleware, and other runtime services. IaaS is the most flexible, but it also requires the user to have more expertise. PaaS simplifies deployment but can reduce the flexibility to customize IT environments. SaaS is even less flexible since it provides the entire infrastructure including user applications. A SaaS user just logs into an up-and-running system. While applications can be configured to some extent, overall, most IT is handled externally. Users may incorporate their business rules but that is about it.

A good analogy relates to car ownership. Gleb from RubyGarage.org breaks it down this way (Gleb 2020):

- *On-Premise Solutions are like owning a car.* You buy a car and are responsible for its maintenance and upkeep. Upgrading means buying a new car.
- *IaaS is like leasing a car.* You lease a car and choose what you want, drive it where you want but the car belongs to the lease company. If you need to upgrade, you lease a different auto.
- *PaaS is like taking a taxi.* You tell the driver where you want to go, and they get you there.
- *SaaS is like going by bus or train.* There is a fixed route, you share the ride and go where you need with everyone else.

Recovery as a Service (RaaS)

Although not in the same category as SaaS, PaaS, and IaaS, some cloud service vendors offer recovery as a service or RaaS. RaaS includes cloud services that facilitate backup, archives, disaster recovery, and business continuity functions. RaaS ensures an organization has data backed up in multiple locations and can quickly resume operations should a natural disaster or other unexpected event occur. In addition to data backups, RaaS may protect and help recover data centers, servers, middleware, databases, web sites, and many other IT resources. RaaS helps businesses reduce downtime and minimize negative impacts on their clients.

Benefits of RaaS Solutions

- Reduce downtime due to data loss or disasters
- Prevent loss of mission critical company data
- Safeguard IT infrastructure and ensure rapid redeployment of resources
- Give cost-effective backup, recovery, and business continuity planning
- Provide geographic-independence of mission-critical resources
- Offer flexibility and multiple backup options

What Are General Benefits of Cloud Services?

Now that we have defined cloud computing and gotten a high-level overview of ways cloud computing is deployed (see Figure 1.7), it is helpful to review and consolidate the advantages this approach to IT offers. Most apply to all forms of cloud computing.

- Cloud service deployment is fast. For many resources, it takes a very little time to configure an application. If the resource is already in place, it may only take a few moments to upsize its capacity or add more seats for new users.
- Cloud services can be accessed from nearly any computing device attached to the internet including smartphones, tablets, laptops, and desktops.
- Cloud services can be accessed from nearly any geographic location and whether someone is in their office, home, or on the road.
- Cloud services are elastic. As an organization grows or shrinks, subscriptions can be increased or decreased to match organizational needs. Cloud providers enable customers to select the level of service that matches their needs.
- Cloud services facilitate improved efficiency and cost reductions. Cloud services are ideal for small organizations, start-ups, temporary projects, and even large organizations wishing to economize.
- Cloud services provide expertise on IT infrastructure without needing in-house staff.
- Cloud services are billed to subscribers, so they pay only for what they use.

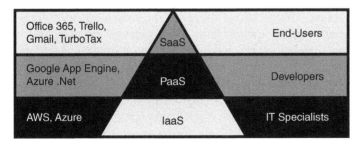

Figure 1.7 Cloud computing and its users.

What Are General Disadvantages of Cloud Services?

Despite the advantages and compelling reasons to utilize cloud computing, several disadvantages do exist. The following points summarize several which will be explored in more detail throughout this book. Among these are:

- Network connectivity considerations. Cloud servers and data storage generally require network connectivity for resource access. If a network disruption occurs, service can be interrupted.
- Cloud security considerations. Whereas cloud computing generally provides high levels of secure access and data protection, an entirely different set of complex issues emerge. For instance, managing access privileges, protecting secure data, and many other topics become important due to cloud computing's nature.
- Cloud services can be costly. Although upfront investment costs may be lower, cloud computing operating costs can add up. Sometimes using a pay-as-you-go model can result in unexpectedly high billing. It is important to track and manage usage costs.
- Vulnerability to attacks. Cloud services provide an online target for hackers. The scale of damage can be vast if a secure system is compromised.
- Loss of control. In cloud environments, both software resources and data generally are entrusted to a third party. This means that corporate governance policies must be enacted by others which could result in compliance issues.
- Technical problems. If technical problems emerge, the fix might depend on cloud service providers. In organizations using multiple vendors or in those lacking an onsite IT staff, this may become a difficult issue to overcome.

What Is the History Behind Cloud Computing?

Cloud computing did not appear overnight as a sudden invention. Rather, it came about due to gradual changes in technology and the ways computers were used. A historic perspective makes this easier to see.

Mainframe computers provide a good starting point. In the early 1960s commercial mainframes began to appear. These early systems were gigantic, not too reliable, and used large amounts of power. The first systems usually were dedicated to single users and were developed for specific business, government, or scientific tasks. IBM dominated this marketplace and gained fame as the computing platform for businesses. Figure 1.8 shows a mainframe computer from this era.

In the 1970s mainframes become more user-friendly. Instead of solely batch-run tasks often dependent on tapes, punch cards or disks, user-terminals with keyboards and display devices appeared. These computers were time-shared and permitted multiple users. They were more powerful mainframes and used a virtual machine computing model. This meant individuals using the computer were allocated memory, which seemed like a dedicated machine, but was shared memory temporarily addressing their computing requests. Users accessed their virtual machines using thin-clients, usually green or amber screen terminals with a keyboard. The computing power resided on a centralized mainframe, but

Figure 1.8 Mainframe Era computer center. Source: National Center for Toxicological Research (NCTR) which is FDA's internationally recognized research center. (CC BY-SA 2.0). https://www.flickr.com/photos/fdaphotos/7421971854/in/album-72157630240761784/

it did not seem that way to the users, except when other peoples' processing slowed their operations.

In some ways, today's cloud environment resembles mainframe era computing. But, one key element is missing: distribution of users over even broader and larger areas. The cloud makes it possible to have thousands and tens of thousands (or even millions) of machines using the same resource pool. With mainframes, virtualization was driven by lack of computing resources. In the cloud, it is driven more by economies of scale.

The computing world did not jump directly from mainframes to the cloud. An intermediate, technology-driven step occurred. This was precipitated by the appearance of the personal computer. In the 1980s, computing power moved to the desktop. Instead of running tasks from a shared, centralized computing source, applications were developed to run on desktop computers. When these were connected into networks, thick-clients became the norm. Computing power existed both at the user's terminal and on servers where data and applications responded to requests from individual users.

This was known as the client-server era. Mainframes now became back office machines running large corporate transaction processing systems and performing other specialty roles. Although PCs and mainframes were connected by networks, many of these were custom developed and only worked internally to an organization. When networking capability was standardized, suddenly computers on different networks could interact and communicate. In the 1980s, TCP/IP became the protocol of what came to be called the Internet. In the early 1990s, the idea of creating an easy-to-use Internet interface emerged in the form of the World Wide Web (WWW). Hypertext Transfer Protocol (HTTP)

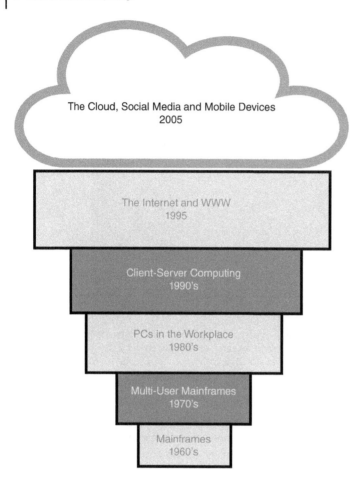

Figure 1.9 General computing eras.

permitted people to access materials stored on web servers without needing to understand the technology.

The WWW was ideal for communication and social media. This came to be known as Web 2.0. Technology companies became clever at distributing software and updates and created online platforms that shared databases among millions of online users. Facebook, Twitter, Instagram, and other social media platforms, enabled by mobile devices, opened computing to even more people.

As in the mainframe days, computing power moved back to shared resources and thin-clients once again became popular. New applications were cloud-based and around 2005, the era of cloud computing, for both individuals and business organizations truly began (see Figure 1.9).

Historic Perspective of Hardware Related to Cloud Computing

Many of the changes driving technology use at both business and individual levels have been brought about by hardware. For instance, the invention of the mainframe opened the door for business computing. The invention of the microprocessor then changed the way computers were built and impacted their sizes. Mobile devices and networking technologies allowed IT specialists to reinvent the way computing services were rolled out.

In general, amazing, and significant changes in hardware capabilities have made the current cloud environment possible. Among these changes were an increase in computing power and low-cost storage capability. For instance, in 1967, a megabyte of storage cost about one million dollars. Currently, that same amount of storage costs less than 2 cents. Similar cost reduction holds true with computing power. In 1985, the Cray-2 Supercomputer was the world's most powerful computer. When the iPad 2 came out in 2011, it offered more power and speed than the Cray-2 for a fraction of the cost.

Network technologies are another area of hardware that have enabled the cloud revolution. Both businesses and individual consumers have access to high-speed broadband. In urban areas, fiber optics make streaming 4K video and other applications possible. Even the most rural locations have satellite Internet capability which offers speeds reasonable enough to access cloud-based resources.

Currently, many cloud providers such as Google, Amazon, Microsoft, Cerner, and others are building and upgrading enormous data centers to power cloud computing. There is no end in sight to the race.

Historic Perspective of Software Related to Cloud Computing

Software has been as important as hardware in the development of cloud-based resources and services. Two major development areas have driven change from the software side. These are: virtualization and SOA.

Virtualization, which came from mainframe environments, commoditizes hardware by taking advantage of its capacity to serve multiple users. VMWare, now owned by Dell Technologies, developed techniques that permitted microprocessors to be virtualized. They found ways to partition and time slice multicore CPUs to enable large numbers of users to share resources. So, when one user was engaged in heavy processing, their application might receive extra resources while another user was demanding less CPU power. Economies of scale allowed better use of existing resources and permitted development of giant, shared server farms.

Operating system virtualization means that software permits hardware resources to run multiple operating system images simultaneously. Each of these virtual operating systems appear real to the user even though they are abstracted away from the actual hardware. In cloud environments, virtualization of servers using a software layer called a hypervisor emulates underlying hardware systems. The emulation may include the CPU's memory, input/output, and networking. Although virtual systems do not run quite as fast as those on actual hardware, for most practical applications, they are fast enough. Greater flexibility and hardware independence offered by virtualization is extremely important in cloud environments.

Another important concept in cloud computing is *Service Oriented Architecture (SOA)*. SOA is a term used to describe the philosophy behind how large development projects can be organized through a divide and conquer and communicate approach. As such, SOA describes both an architectural style and vision of how an organization should approach developing, building, and deploying systems. The main mindset of SOA is to develop reusable services that can be integrated to create large scale systems. The days of building gigantic, specific applications have been replaced with creating a set of building blocks used to perform specific functions. SOA is an enterprise-level approach to development and is not meant for single application development.

SOA: A New Way of Doing Old Things

- Code developed in reusable modules that can "talk" to each other.
- Encapsulate design decisions that may change in the future (e.g. all changes can be made in a single, known location).
- Ensure code modules can be combined in different ways depending on specific software deployment needs.
- Draws from fundamental, object-oriented, software design principles.

In many regards, SOA has rebranded software development principles forming the foundation for software engineering practices. A few differences do make SOA unique from the design environments of the 1980s and 1990s. First, SOA is meant to work in networked environments. Software modules "talk" to each other and communicate over a network rather than through traditional procedure calls or parameter passing. SOA uses messages to request responses or provide updates. Another difference is in variable declarations and use. In most cases, SOA architecture does not rely on global variables for tying modules together. If global style variables do exist, the values reside in a database that has been configured to control concurrent access and ensure multiple clients can update or read the values without interfering with each other.

So, SOA can be described as consumer components receiving services from provider components using an infrastructure provided by cloud computing. It is helpful to use an analogy to describe a complicated topic. Many IT consultants use Legos to describe SOA. Here is a version of this analogy:

SOA Explained in Terms of Lego[1] Blocks

From our brief description of SOA, we know that this concept means that software services should be built in ways that are reusable, flexible, and independent. This is like Lego blocks. Here are a few characteristics:

- *Legos are interoperable.* Legos have standard bumps that can fit into any other Lego block. Having a standard interface means that even when new block styles are designed, they can still plug directly into the existing structure without an issue. In SOA, the bumps are the messages that link modules together.
- *Legos are composable.* A single Lego block may be interesting to look at, but it does not do much. Until a structure comprising multiple Legos is created, not much value exists.

---

1 *LEGO*® *is a trademark of the* LEGO Group of companies.

- *Legos are reusable.* A person can build a structure with Legos, then later the same blocks can be reassembled to build something different. The blocks are not wasted, just reused. Some IT leaders like to extend the metaphor and say different colored blocks represent different services. For instance, yellow blocks might represent marketing services and blue blocks accounting services. This enables the composition of different organizational systems based on current business rules.
- *Legos are robust.* Although breaking individual blocks is possible, the blocks remain unlikely to break when structures are reassembled. That same feature applies to SOA modules. They should be developed, tested, and quality approved before ever being used in software development. SOA components must be designed for robustness.

So, what does the Lego analogy mean for businesses? It really represents what organizations want from their IT systems. They do not want to know what is inside a block, but they do want to ensure the blocks perform as needed. They must be interoperable, composable, reusable, and robust. The characteristics of Legos make them flexible and responsive. Software should be the same.

Of course, several downsides do exist in the Lego analogy. First, Lego structures are not glued together and can fall apart. When a Lego structure is complex, it becomes more difficult to manage. It is the same with software systems. Legos link well to other Legos but not to toys from other manufacturers. Legos come in many different sizes. Small blocks are more flexible and permit custom projects but take longer to assemble and require managing more pieces. Larger and specialty blocks make construction faster but offer less flexibility. It is the same with business services. There must be a balance between flexibility, reuse, and practicality. Granularity is a term used to describe the level of service of software components.

Harriet Fryman, the senior director of product marketing for Cognos, provided these seven principles of SOA during an interview with Loraine Lawson published on ITBusinessEdge.com (Lawson 2007).

1) Open and standards based.
2) Platform-neutral. Encapsulated so a service works identically on Linux or Windows.
3) Location-transparent. Should not change based on user or location in the global infrastructure.
4) Peer-to-peer. There should be no primary or secondary (Landau 2020). Every service is created equal. This means services can spawn and scale out without a single point of failure.
5) Loosely coupled. Can enhance capability within a service without impacting another service.
6) Interface-based. Each service is opaque regarding other services.
7) Coarsely grained. Must operate at a business level, not down in the bits and bytes. That makes it more reusable.

The Lego analogy is helpful but not a complete description of SOA. Remember, SOA architecture relies on components "talking" across networks and sending requested data back as a service. Legos physically link together. In either case, having robust, composable, reusable and interoperable services is a worthwhile goal for any organization's software services.

Use of SOA requires both technological and organizational mindset changes. While services become easier, other concerns, like security issues emerge. SOA is the logical extension of the browser-based standardization that was responsible for the emergence of the WWW. When these principles are used with machine-to-machine communication, standardization can emerge. SOA makes it possible to compose independent services into business applications, and this is particularly well suited to make SOA the de-facto architectural model for building virtualized applications running in cloud-based environments (Lawson 2007).

Summary

Chapter 1 has given us an overview of cloud computing and primary concepts that have enabled it to become the leading approach for rolling out enterprise IT resources. Cloud computing is an ideal solution space in the current mobile computing environment. Users have access to remotely hosted data, services, applications, storage, and systems. In the most general sense, cloud computing is a delivery system like a public utility. Clouds can be privately deployed with all resources hosted internally to an enterprise using cloud technologies (e.g. an intranet), publicly from large server farms meant to develop economies of scale by sharing overhead management costs, or in hybrid solutions that retain some data and services on premise while others are hosted off-site. See Table 1.1 for main cloud computing features.

Three main operational paradigms are common in cloud computing. These are:

- SaaS (software as a service)
- PaaS (platform as a service)
- IaaS (infrastructure as a service)

With SaaS, hardware, operating system, and applications are supplied and managed by an external party. Everything but the applications are transparent to the user. With PaaS, hardware and operating system are supplied and managed by an external party. The hardware functions are transparent to the user, but the users supply their own applications. Finally, with IaaS, hardware is supplied and managed by an external party, but the operating systems and applications are managed by the user organization.

Cloud computing delivers storage and applications as a service, rather than a product, offering both cost and business advantages. To do this, cloud computing moves services off-site to a hosting company, or a centralized self-hosted facility. Centralization accomplishes what IT managers seek: increased computing capabilities without having to provide new infrastructure. Cloud capabilities include rapid deployment, scalability, elasticity, access to IT expertise, and cost reductions.

Cloud computing uses the concepts of virtualization and SOA to create and manage resources. Virtualization leverages hardware capabilities by ensuring resources are dynamically allocated to those needing the capacity at the correct times. VM software partitions a physical computing device into multiple virtual devices, which can be used and managed to perform organizational computing tasks. Virtualization offers agility and can speed up IT operations and increase infrastructure utilization. SOA is a philosophy for software

Table 1.1 Cloud computing primary features.

Cloud computing feature	Description
Pooled resources	Available to users with subscription and level of access.
Virtualization	Improved utilization of hardware assets.
Elasticity	Dynamic scaling capability (e.g. scalability) and the ability to downsize when needed.
Automation	Build, configure, and deploy without human intervention required.
Metered billing	Per-usage business model. Pay only for use.

development and deployment where independent, reusable services are developed in ways that permit integration with large scale systems.

The possible uses of cloud computing are plentiful. Users interface with IT resources through their web browser, eliminating the need for installing numerous software applications. Organizational applications interface through cloud services to access resources and exchange information in cost effective and manageable ways.

References

Gleb, B. (2020). IaaS vs PaaS vs SaaS. https://rubygarage.org/blog/iaas-vs-paas-vs-saas (accessed 7 July 2020).

Landau, E. (2020). Tech confronts its use of the labels 'Master' and 'Slave'. https://www.wired.com/story/tech-confronts-use-labels-master-slave (accessed 9 July 2020).

Lawson, L. (2007). The merits and seven principles of SOA. https://www.itbusinessedge.com/cm/community/features/interviews/blog/the-merits-and-seven-principles-of-soa.

Bibliography

Alam, T. (2020). Cloud computing and its role in the information technology. *IAIC Transactions on Sustainable Digital Innovation (ITSDI)* 1 (2): 108–115.

Armbrust, M., Fox, A., Griffith, R. et al. (2010). A view of cloud computing. *Communications of the ACM* 53 (4): 50–58.

Buyya, R., Broberg, J., and Goscinski, A.M. (2010). *Cloud Computing: Principles and Paradigms*. Wiley.

Hayes, B. (2008). Cloud computing. *Communications of the ACM* 51 (7): 9–11.

Hurwitz, J.S. and Kirsch, D. (2020). *Cloud Computing for Dummies*. Wiley.

Jamsa, K. (2012). *Cloud Computing: SaaS, PaaS, IaaS, Virtualization, Business Models, Mobile, Security and More*. Jones & Bartlett Publishers.

Mell, P. and Grance, T. (2011). The NIST definition of cloud computing: special Publication 800-145. Gaithersburg, MD. https://csrc.nist.gov/publications/detail/sp/800-145/final.

Rani, D. and Ranjan, R.K. (2014). A comparative study of SaaS, PaaS and IaaS in cloud computing. *International Journal of Advanced Research in Computer Science and Software Engineering* 4 (6).

Singh, A. et al. (2016). Overview of PaaS and SaaS and its application in cloud computing. In: *2016 International Conference on Innovation and Challenges in Cyber Security (ICICCS-INBUSH)*, vol. 2016, 172–176.

Velte, T., Velte, A., and Elsenpeter, R. (2009). *Cloud Computing, a Practical Approach.* McGraw-Hill, Inc.

2

Who Uses the Cloud?

Cloud computing offers major benefits across a wide spectrum of users including people operating at the individual level, employees of small and medium businesses and those in corporate environments. Chapter 2 provides insight into various aspects of cloud computing that appeal to each of these. Unique benefits across so many levels ensure cloud computing has become a paradigm-shifting, game-changer. This chapter focuses on how the cloud is used by various groups and demonstrates that this technology is more than a passing fad.

Individuals Users

An exciting trend in cloud computing relates to individual users. Personal storage of digital resources has grown tremendously in the past five years. People want to store their photos and videos, movies, music collections, eBooks, documents, family records, recorded television programs, digitized art, souvenirs, digital keepsakes, correspondence, text message streams, video diaries, and countless other artifacts. Managing and keeping these items secure has become important. As a result, many individual users have turned to the cloud. See Figure 2.1.

Individuals are motivated to use online storage for several reasons. For one, the idea of having all digital resources handily stored in a single place is appealing. Another reason relates to a sad reality: it is very easy for digital resources to disappear without the possibility of recovery. A laptop can be stolen; a hard drive might crash; a tablet device might be dropped and broken; or a power surge could ruin a desktop computer. Cloud storage separates data from fragile digital devices and safeguards our increasingly valuable holdings.

The cloud also provides access to software and services for individuals. Software subscriptions provide tools ranging from Microsoft 365 to an enormous variety of mobile apps, online games, social media platforms and daily utilities (search engines, navigation aids, and so forth). Individuals enjoy access to low cost, easy to maintain and deploy software due to the cloud. Without a doubt, the cloud is the primary mode of software acquisition and use for most people (Figure 2.2).

Individual users also benefit from cloud storage. The next sections of this chapter focus on three primary cloud solutions for individual users. The first solution is a subscription to a cloud storage solution like Dropbox, Box.net or one from other vendors offering online storage services. A second option, one that is becoming more popular in recent years, is

Cloud Technologies: An Overview of Cloud Computing Technologies for Managers, First Edition. Roger McHaney.
© 2021 John Wiley & Sons Ltd. Published 2021 by John Wiley & Sons Ltd.
Companion website: www.wiley.com/go/mchaney/cloudtechnologies

Figure 2.1 Moving a CD music collection to the cloud.

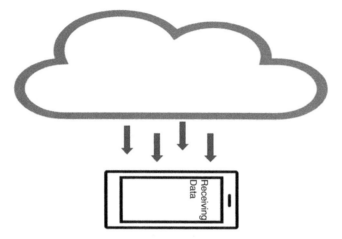

Figure 2.2 The cloud delivers software to individual users.

called personal cloud storage (PCS). This means the individual user invests in cloud drive hardware, sets up, and manages that solution. The third approach is to work with a vendor that offers a personal storage cloud (unlike Dropbox-type vendors that let you become part of their multiuser cloud). In all cases, the individual user can consolidate their files from multiple personal devices such as their phones, laptops, desktops, Smart TVs, video security devices, game consoles and so forth into a location accessible via the Internet. Next, we will look at these three models in more detail: public cloud subscription, PCS, and personal hosted solutions.

Public Cloud Subscription Storage for Individuals

The first move by many end-users is to subscribe to a public cloud storage service because it is inexpensive (often free for smaller amounts of storage). Examples of these services include Box.net, Dropbox, Google Drive, Carbonite, and Apple iCloud. An advantage to these subscription services is that no upfront investment is required. In most instances, the user creates an account, logs in and uploads files where they remain until the user decides to download them for use later. The host providers update software, perform backups, maintain security, and offer other services such as file sharing, use statistics, and identification of duplicate files. Users are responsible for not overwriting their files (although some vendors offer features to recover lost or earlier versions of files), managing

Figure 2.3   Box.net cloud storage user interface.

the sharing permissions, and, in some cases, paying a subscription fee. Most services also provide features to automatically store or sync files between personal devices like smart phones and the storage platform. Figure 2.3 shows the interface of Box.net.

One concern about using public clouds relates to privacy. While data is secure and maintained carefully, privacy may be an issue. For instance, sensitive financial data, health information, and personal data may be stored. Several public cloud hosts strive to minimize privacy issues with zero-knowledge, end-to-end encryption software solutions. Zero-knowledge encryption means the data owner has complete control of the cloud encryption keys. The host cannot decrypt or see the content of the files in any meaningful way. Only the data owner can decrypt the files. This eliminates government surveillance, unwanted intrusion into private financial records or photos –or whatever the owner wishes to keep secure and private. In most cases, the files are encrypted on the local machine prior to being sent to the cloud and only decrypted upon return. One cloud vendor, sync.com, has made this their niche by marketing Zero-knowledge encrypted cloud storage. We will cover encryption in cloud services in more detail in chapter 8.

Private Cloud Storage (PCS) for Individuals

The second option for individual cloud storage moves beyond services offered by public subscriptions. PCS enables the user to do the same things that public cloud subscription services do, including syncing devices, sharing files, and accessing content from smart phones or other devices. But a big difference exists: the individual owns the hardware.

Having a personal or private cloud makes sense for a few different reasons. First, a primary benefit is speed. If someone uses their own hardware, they can expect to be able to synchronize between devices quickly and can stream media faster and more reliably. Self-hosted clouds can be physically closer (for instance, a person can keep their "cloud" device right next to their smart television in their home). The data is not subject to potential internet traffic congestion and uses local WiFi for a connection instead. Another speed-related advantage is that many PCS devices permit the user to plug in directly. In a situation where the Internet is down, it is still possible to directly connect to the data (Figure 2.4).

A **home media server** can be set up using any storage on an existing computer. Once installed, Plex Media Server software can be configured to access music, video clips,

(Continued)

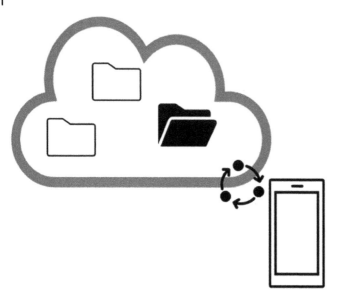

Figure 2.4 PCS enables automatic synchronization of personal devices from anywhere.

Figure 2.5 PLEX.

(Continued)

movies, and other media from networked computers or hard drives and then makes it available for networked home devices. For many users, this means their smart TV. See Figure 2.5.

Other PCS benefits relate to privacy. A PCS device is within the custody and control of the owner. While services like Google Drive are secure and safe, they are subject to governmental regulations and Google may share your data if required. A PCS device can be removed from service and data erased if desired. The data also can be duplicated more easily and moved to other devices. In some online storage venues, this is very difficult to accomplish. The PCS owner can use encryption software and retain control of their key to ensure higher levels of privacy.

PCS also eliminates recurring costs. Subscription models for storage require regular payment. Increases in storage capacity incur more cost. A PCS system is reasonably priced and can be quickly put together. A good system can include additional drive bays and USB or other connectivity options. This means storage can be added when needed.

Several cautions do exist regarding PCS. For instance, the PCS device itself is susceptible to damage from a variety of sources: fire, flood, tornado, electric surges, and theft of the physical device. Likewise, no automatic off-premise backup is maintained unless special measures are taken. Another downside to using a PCS device, is that upload speeds can be poor when remotely accessing a home network where the device is attached. Many experts recommend using a hybrid approach. This means using a PCS and then subscribing to a backup service like Sync.com to ensure your cloud's redundancy.

Most PCSs use a local network-attached storage (NAS) approach. These devices have their own CPU, memory, and operating system. They have an ethernet connection to plug into a router for access via WiFi. NAS devices can have single or dual drives. NAS devices generally are configured through a software interface viewed using a Web browser. NAS multidisc devices may provide RAID (redundant array of independent disks) or JBOD (Just a Bunch of Disks) striping capabilities. These technologies permit adding capacity when required.

RAID (redundant array of independent disks) is a technology that redundantly stores data in different places on multiple hard disks to protect data in the case of a drive failure. Think of it as a safety measure internal to the device to protect against partial failures. Data is kept safer using this technology.

Several options exist for purchasing PCS NAS devices. Some vendors offer specialized versions of these devices that are optimized for video or music streaming with easy connections to home media centers. Among the leading vendors and products in this area are: Western Digital (WD) My Cloud Personal, Seagate Technology Personal Cloud, Buffalo Americas LinkStation, LaCie CloudBox, Synology, Inc. DiskStation, and QNAP Systems TS-x51+ series. Many of these vendors offer multiple cloud solutions which may include single or dual drive versions. Speeds, capacities, and prices change continually with a trend toward faster devices and more capacity at lower prices.

Virtual Server Software

Cloud solutions can be implemented with software and existing hardware. For instance, OwnCloud offers server software that a user installs on their own computer. Client software is loaded on all their mobile devices and computers to permit accessing this newly configured virtual cloud drive from the Internet. Rather than buy new hardware specifically designed for cloud use, this DIY approach to cloud storage makes sense for budget conscious end-users with reasonable tech skills.

Hosted Personal Cloud Storage Using Third Party Hardware

A third option for creating a personal cloud involves using a hosting service where cloud software can be installed and deployed. In most instances, this requires finding a web host that offers preinstalled cloud software instances or finding one that has the infrastructure necessary to install cloud software such as OwnCloud. Usually, this means the hosting service supports PHP and MySQL, or other software. Smaller companies like Bluehost and Dreamhost offer these services. It is also possible to have Amazon Web Services host Own-Cloud or similar software.

Following selection of a host, the cloud software needs to be installed. Acquiring a URL to easily access the files remotely is also helpful. Setting up a cloud server with a third-party host offers advantages comparable to public cloud providers – the data are usually stored in secure, climate-controlled facilities. The data are kept across multiple servers in different geographic locations. The data are backed-up and multiple copies are maintained.

Privacy can be another advantage when using a hosting service. Software can be used to encrypt files before sending them to a host. As mentioned earlier, this is *zero-knowledge encryption*. Files are encrypted before leaving the owner's machine and are not decrypted until returned from the cloud to the owner. Only the owner knows the decryption key. Software packages such as nCrypted Cloud, Boxcryptor, and Sookasa provide these services.

Public Cloud versus Personal Cloud Storage

The cloud offers new options for personal data storage. Depending on user goals, it might be better to consider a large subscription service that is relatively worry-free, or it might be better to use a personal cloud offering better security and more control of the data. A hosted personal cloud is the third option offering benefits found in both personal and public subscription storage clouds. Table 2.1 and Figure 2.6 provide a summary of the pros and cons of each approach.

Small and Medium Enterprise (SME) Users

Most benefits enjoyed by individual users also apply to small and medium enterprise (SME) users. SMEs have embraced cloud computing because it makes great sense to an organization that wants to save money, focus on its core mission rather than maintain IT infrastructure, wants the capability to scale up quickly with growth, and needs to have access to the latest technologies. SMEs can realize increased profitability due to agile working processes, task automation, real time information updates, collaborative computing tools, and access to customers via social media. This is particularly true for entrepreneurs and start-ups that need to quickly move without upfront investments.

Simply put, cloud computing gives SMEs access to business applications and data from anywhere at any time from all Internet-connected devices, at a reasonable price. SMEs can use tools previously only available to large companies including accounting software, CRM systems, ERP software, HR systems and so forth. Cloud software vendors offer subscription-based models that enable small companies to pay lower costs and expand services as they grow. For instance, invoicing, payroll systems, employee scheduling, HR services, contracts, and legal document services – all can be acquired via online subscription-based services.

How Can Cloud Computing Save SMEs Money?

Cloud computing can save money for SMEs in several ways. Many of these resemble the benefits we discussed when describing individual users. Included are:

- *SMEs do not have to pay an internal team of IT staff members for many routine tasks.* For example, no one needs to be on-premise to install and update software, manage

Table 2.1 Pros and cons of various cloud storage attributes.

Public cloud subscription	Personal cloud	Hosted personal cloud
Cost considerations		
Monthly charges	No monthly charges	Monthly charges
Data loss potential for non-payment	No potential for non-payment	Data loss potential for non-payment
Storage costs decline each year	Storage costs decline each year	Storage costs decline each year
Very low-cost initial investment	Initial equipment costs	Low cost initial investment
Adding capacity easy	Adding capacity requires equipment purchase	Adding capacity is easy
Security considerations		
Hardware secure	Hardware can be stolen or damaged	Hardware secure
Subject to government surveillance	Easier to protect contents against government surveillance	Subject to government surveillance
May be targeted by large scale data theft by hackers	Of little interest to hackers	Not likely to be a hacker target
Encryption requires more work on user side for most providers (exceptions like Sync.com exist)	Can be encrypted securely for full privacy	Can be encrypted but with some end-user knowledge required
Data is not under full control of end-user	Data under full control of end-user	Data not under full control of end-user
Host may collect data regarding use	No data about use collected by external parties	Unlikely that user meta-data will be collected
No need to configure equipment or software	Must configure software and hardware (e.g. router and NAS device)	Must configure software
Sharing		
Collaboration easy	Some difficulties sharing data	Collaboration requires more overhead for managing users
Hard to move data to other services	Easier to move data	Hard to move data to other services
Difficult to clone data	Easy to clone data	Somewhat difficult to clone data
Backup and recovery		
Secure backups	Must create own backups	Secure backups
Less prone to large scale disasters	Less secure during large scale disasters	Less prone to large scale disasters
Hardware failure not a concern	Hardware may fail	Hardware failure not a big concern
No access without Internet connection	Access without Internet connection	No access without Internet connection
Vendor may go out of business	Less worries about vendor	Vendor may go out of business

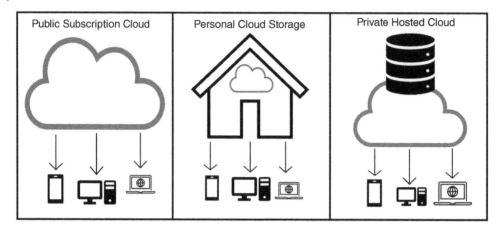

Figure 2.6 Three hosting solutions for cloud storage.

Figure 2.7 Ensuring the human element with SaaS.

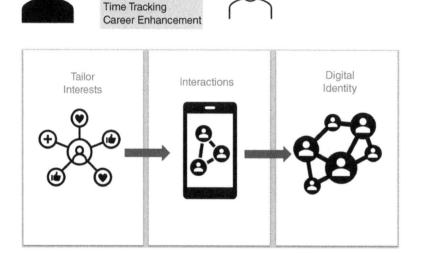

Figure 2.8 Social media can be tailored to provide an SME with a digital identity.

servers, handle email accounts, run backups, and maintain security. Many of these duties become the responsibilities of the cloud vendor. Depending on the SME's mission and needs, different levels of service can be employed, ranging from IT being completely managed off-premise with a high level of service to situations where employees share online resources and are subscribed to SaaS services similarly to how individuals use these tools.

- *Software costs are reduced.* Organizations can subscribe to software like Microsoft 365 and Trello to enable their employees to interact in the cloud. There is no need to purchase

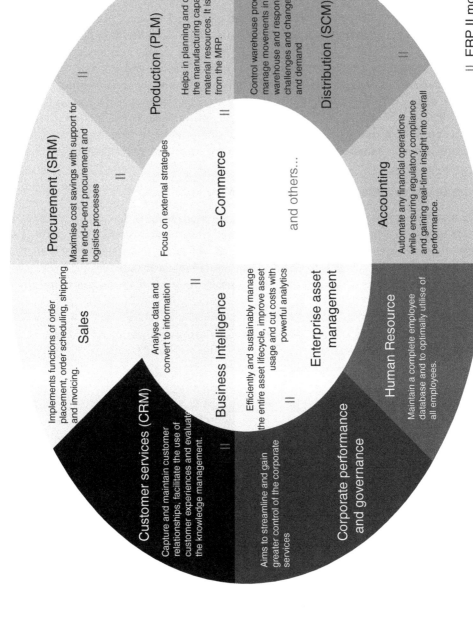

Figure 2.9 ERP software functions. Source: Shing Hin Yeung, Enterprise systems modules, 2013. Licensed under CC BY SA 3.0.

Procurement (SRM)

Maximise cost savings with support for the end-to-end procurement and logistics processes

Production (PLM)

Helps in planning and optimising the manufacturing capacity and material resources. It is evolved from the MRP.

Distribution (SCM)

Control warehouse processes and manage movements in the warehouse and respond faster to challenges and changes in supply and demand

Focus on external strategies

e-Commerce

and others...

Accounting

Automate any financial operations while ensuring regulatory compliance and gaining real-time insight into overall performance.

Implements functions of order placement, order scheduling, shipping and invoicing.

Sales

Analyse data and convert to information

Business Intelligence

Efficiently and sustainably manage the entire asset lifecycle, improve asset usage and cut costs with powerful analytics

Enterprise asset management

Human Resource

Maintain a complete employee database and to optimally utilise of all employees.

Customer services (CRM)

Capture and maintain customer relationships, facilitate the use of customer experiences and evaluate the knowledge management.

Aims to streamline and gain greater control of the corporate services

Corporate performance and governance

= ERP II modules

software and install it on the SME's servers/computers. Cloud applications can be found for most software functions.

- *Hardware costs are reduced.* Servers and computer hardware costs can be reduced dramatically with cloud-based services. SMEs using IaaS or PaaS can use cloud storage, virtual servers, and enjoy the benefits of having data backup, physical security, updates, and recovery services managed off-site.
- *Storage costs are reduced.* Like individual users, SMEs can use cloud resources for data, document, and information storage. For example, sensitive data can be stored on privately managed internal NAS cloud devices. Other data can be stored in subscription services provided by companies like Box.net which have specific business solution packages. Other options exist for setting up hosted provider clouds and servers.
- *Integration across organizational applications is made easier and less expensive.* An SME might be able to find cloud computing services to enable operations that previously were too expensive for small enterprises. For example, software like SAP or JD Edwards automated many organizational computing needs with a single corporate wide database. A variety of vendors supply similar enterprise software using cloud technologies.

What Cloud Computing Features Appeal to SMEs?

Cloud computing applications appeal to SMEs for more reasons than just cost savings. Cloud applications usually are browser-based, and this means business users can access software and resources using mobile devices such as tablets and smartphones. For example, a person running their consulting business can use a mobile device to bill and accept payment while at a client's worksite. Traveling sales staff can track their expenses in real time while on the road.

Another area that appeals to SMEs is the cloud's scalability. As a business grows, its IT infrastructure also can grow. An SME can create long term growth plans that accurately predict costs of scaling up. And, benefits of powerful cloud-based business software can be used without needing a large upfront investment.

SME Cloud Software

For SMEs, many benefits of cloud computing relate to the software applications available. Most traditional business computing can be done with a SaaS model rather than through desktop software installation. The next section of this chapter briefly reviews popular business application areas to provide a sense of the amazing growth and high impact of cloud-based application software for small and medium sized business. Many of these applications also are used by individuals.

Accounting Software

Accounting software has been synonymous with business for decades. Recently, cloud-technologies have encouraged desktop accounting software vendors to migrate to cloud-based solutions. Businesses have responded and many SMEs manage accounts payable, accounts receivable, payroll, general ledger, financial reporting, maintaining controls, and annual tax returns using SaaS. Under this model, SMEs pay a monthly fee to use services. Many vendors provide these with a menu type approach where the organization pays to subscribe to a subset of available services.

In addition to the ease of using accounting software, many best practices are built into the software. Legal requirements, international conventions, reminders, educational materials, access to user groups, and other benefits come with the software subscriptions. Legal storage requirements for financial records and security backups can reduce compliance burdens on small companies that do not have staff accountants. Many SaaS accounting applications provide mobile apps which make receiving payment, paying bills, checking account use, and tracking expenses much easier and more agile. The following vendors are leaders in online accounting systems:

QuickBooks: QuickBooks is an accounting package developed by Intuit, who also offers TurboTax and other software. QuickBooks is a fully featured accounting package offering full reporting, decision support, tracking, and forecasting capabilities. QuickBooks is very popular and is a leading online accounting software package in the USA.

Xero: This software provides a wide range of accounting services including invoicing, inventory, payroll, budgeting, and expense tracking. Xero offers tools to import an SME's banking, credit card, and PayPal data. Xero also offers a popular mobile app.

Zoho Books: This is a highly customizable software package that integrates with Zoho's broader suite of office products. Zoho books is more of a bookkeeping software package and issues invoices, sorts expenses, and provides inventory management features. It can connect to bank accounts. Zoho does not focus on payroll or other accounting functions but continually adds new business options to its portfolio. It is a low-cost provider.

Sage One: This is an accounting package for small businesses offered by a large ERP company. It is fully featured and is built on a powerful database system that can integrate with other business functions.

Information Is the Main Difference Between Financial and Managerial Accounting

Most SaaS accounting software packages offer both managerial accounting and financial accounting features. Managerial accounting focuses on data collection and processing into information reported to internal managers and executives to support planning, controlling, and decision-making. Financial accounting is focused on collecting and reporting information for external users like investors, creditors, regulators, bankers, and auditors.

Human Resources (HR) Software

Human Resources (HR) SaaS applications have made it possible for a SME to outsource many of its routine employee interactions. Among these functions are storing and updating employee data, payroll, vacation tracking, sick or personal day tracking, training records, certification, recruitment, benefits administration, attendance records, evaluation, legal requirements, and performance management tracking. HR systems can automate processes and rely on best practice knowledge developed over many years by specialists in these areas.

HR software tools take a database-driven approach to enable SME owners or managers to view, compare, and compile reports from employee records. Often these tools provide an interface with portions used by the manager and other portions used by the employee. For instance, vacation requests, evaluation results, training requirements, and other items have both a manager's view perspective and an employee's view perspective. Many HR packages also tie into accounting or other organizational software platforms. Good HR software packages enhance organizational relationships rather than replace them. People need to feel connected and the software should facilitate that, not remove the human connections within an organization. See Figure 2.7.

Especially important capabilities for most HR software systems usually feature:

- *Benefits Administration*: This includes enrollment options for insurance, retirement plans, long term disability, family leave, and so forth. HR software must be customizable to provide the options in a meaningful and easy-to-access way for both managers and employees.
- *Shift Management*: In businesses where employee coverage of activities is important, HR software can provide tools for managing shifts including rules for scheduling, employee managed coverage, trading shifts, and vacation planning. Some systems offer mobile options so on-call workers can be texted or statuses updated to ensure smooth transitions between shifts, and to facilitate situations where worker capacity demands need dynamic management.
- *Application Tracking*: This includes the ability to manage job postings, track applicants, manage candidate interviews, track applicant source effectiveness, and employee onboarding.
- *Performance Management*: This ensures both employee and manager keep a record of employee goals and provides ways of tracking goal accomplishment. Some packages tie this to compensation and payroll.
- *Training*: This can be a separate module that tracks employee training goals, certifications, and completions. This may be crucial where compliance is an important consideration.
- *Legal:* Having a paper trail regarding infractions and rule violations may be an important aspect of an HR system to provide legal safeguards for management and an organization. Employee hiring and firing falls into this area.
- *Other Helpful Tools*: Many HR managers find themselves writing employee handbooks, job descriptions, preparing company hierarchies, and developing legally binding documents. Several cloud-based HR platforms do a great job with these features. A company needing any of these items specifically should be sure to shop around to find a package that offers the right fit.

The following vendors are leaders in the online HR software area:

BambooHR: This package offers an appealing, modern interface. It covers all basic HR functions including a rapid employee performance management module. It has an open API to allow connections to other cloud-based software packages. Some critics say its benefits administration functionality needs improvement.

Zenefits: This "people-based" platform is easy to use and intuitive with fully featured benefits administration options. It integrates with payroll systems from most major vendors including QuickBooks. While its base package is reasonably priced, its add-ons can rapidly add cost.

Namely: This is a highly customizable package with an outstanding user interface. Unfortunately for smaller businesses and startups, it is expensive and takes time to initially set up. It also has a costly, fee-based payroll and benefits administration module. Many competitors include these features as part of their main package. Namely has excellent HR functionality.

SAGE: Like its accounting package, SAGE HR is highly configurable and made to handle more complex businesses with multiple divisions or branches in overseas locations. It moves the user into the realm of ERP with applicant tracking, benefits administration, and integration with other SAGE tools. Additionally, it easily integrates with third-party payroll and other packages. It is built on the Salesforce.com technology platform. This presents both pros and cons depending on whether an organization plans to use Salesforce.com or not. It is highly configurable.

SAP SuccessFactors: Many consider this the best HR cloud software for SMEs with setup wizards, video tutorials, and other multiple help features. Since it is the SME offering by the world's largest ERP/Enterprise software company, it is highly scalable to other SAP products should an organization need to rapidly grow. It is good choice for a startup that foresees itself being acquired by a larger company in the future since SuccessFactors can be easily integrated with SAP systems running in other organizations. On the downside, this is not a full ERP system and it is more expensive than most of its rival software. On the plus side, the customer support for this package is unrivaled. SAP uses the paradigm of human experience management (HXM) as a theme for this software.

EffortlessHR: This platform was created by a team of HR professionals and so it contains solutions for the most time-consuming HR tasks. Some critics consider its interface dated, but the software is very powerful and fully featured. In addition to primary HR functions, it provides tools to write employee handbooks, set up an employee time clock, conduct and track employee assessments, and build job descriptions. It is an excellent package for a unionized environment. It also provides a wealth of business forms that can save an organization time and money. It is a reasonably priced package with great functionality.

An SME should consider a few attributes before committing to an HR system:

- What are the degrees of system flexibility, elasticity, and scalability?
- If needed, can the software import data from spreadsheets, databases, or paper documents?
- Does the system have an API to permit integration with other systems and data?
- Can the system accommodate the organization's business rules?
- Will the system grow and scale with the organization?
- If payroll and other functions are provided by third party vendors, can information be easily shared?

(Continued)

(Continued)

- What level of support and training does the vendor offer?
- Who will own the system-managed data, and can it be downloaded if the software vendor goes out of business?
- What security measures are built into the HR information technology?
- Will employees be able to enroll in benefits plans and make changes in real-time to their personal data and plan choices?
- Will the technology provide Human Resources with the authority to customize access to functions and data?
- What types of reporting capabilities are available from the HR information technology?
- Can reports offer multiple views and formats?
- Do drill down features exist to enhance decision-making?
- Does the software offer specific features for employee communication?
- What is the vendor's customer service record?
- What are the system costs (initial and long-term)?

TidyForm and Spreadsheet Templates: For the company that is essentially being run as a Mom and Pop shop (e.g. very small business with few employees), the cloud-based website, TidyForm offers documents that can be used in lieu of a fully featured HR system. This includes employee contracts and other needed items. Add a "Sheet" from Google docs for tracking and a small business can have an operational, although basic, system to help maintain HR records.

In general, human resource information systems (HRISs) are essential for companies to manage benefits plans and employee information. This has become more possible with SaaS cloud platforms. The industry is currently in a period of innovation and competition. Benefits management technology is a key element in HRISs to help manage information and track money spent on benefits plans. The result is a database of key employee metrics that can be used in organizational decision making.

Customer Relationship Management (CRM)

Customer Relationship Management (CRM) started as a strategic approach for interacting with customers in a way that could be tracked and maintained. The goal was to build a relationship with a client and make sure the relationship was reinforced positively through each contact, even if different organizational members made the contact. CRMs keep customers connected and provide a consistent message to a client from all parts of the organization. CRM was revolutionized by Internet-based communication capabilities which form the basis for many cloud-based CRM tools.

Although CRM started as an approach/strategy for organizations, the term has come to mean CRM systems, processes or software packages that help track an organization's relationships and interactions with current customers and potential customers. The software helps companies stay connected to customers, receive reminders about contact, streamline interaction processes, improve business relationship, enhance communication, and ultimately create a path to greater profitability. CRM software has expanded to include relationship management with traditional customers, service users, colleagues, suppliers, and a variety of other stakeholders.

At its core, CRM software is used by an organization to record customer contact information such as email, Twitter handle, telephone numbers, social media profile, Website address, and any other details that could be helpful. Many CRM systems are configured to pull data in from social media feeds, news services, and other sources to create more complete pictures of stakeholders, partners, clients, or prospective clients. CRMs also store personal details about clients or stakeholders to enable richer conversations. These details might include birthdays, the content of prior conversations recorded by colleagues, or client's communication contact preferences.

The CRM system organizes data into a complete picture of the stakeholder that helps track and build relationships over time. Many CRMs have a goal of creating a 360° view of the stakeholder with both business interaction and superficial information needed to conduct a more engaged conversation. For instance, imagine that a customer mentioned they were about to run in a marathon. This could be recorded so during a subsequent contact the customer could be asked how their marathon turned out.

Good CRM software not only tracks key business and personal information, it also provides a customizable dashboard with an easy-to-see interface used during a customer discussion by phone or that is reviewed prior to a face-to-face meeting. Ideally, everything is consolidated and provides a view of the stakeholder's previous history, order status, outstanding and resolved service requests, and bits of personal information to enrich the conversation.

CRM Benefits

- Enhanced stakeholder contact management
- More effective cross-team interaction and collaboration
- Improved sales forecasts
- Tools for lead management
- Better customer relationships
- Consistent handling of stakeholders
- Increased productivity
- Richer report and decision-making capability
- Improved customer retention

CRM systems traditionally have been used as sales and marketing tools but also are common in customer service, supply chain management, and partner management systems. Sales representatives and managers have used CRM to better predict and manage their sales pipeline. This means understanding customer patterns and using these to facilitate the sales process. In addition, CRM software is used by sales managers to understand the processes and progress of individual sales team members in relation to goals or targets. CRMs provide goal management, product tracking, and campaign management. Best practice approaches are built into most CRMs. For example, marketing specialists use CRM-gathered data to forecast demand and view historic trends. Systems collect then correlate social media chatter with business demand and sales pipeline developments (Cheng and Shiu 2019).

CRMs enable a concerted effort by an organization. For instance, if one team member in sales speaks with a customer and that same customer later speaks to a customer service team member, a consistent message is more easily maintained. Likewise, customers may communicate using different channels: email, Facebook, live chat, Twitter, or phone. The CRM can help ensure access to all communications no matter how it was presented. This is particularly helpful for customer service specialists that may receive help requests in multiple ways.

In the supply chain area, CRMs facilitate effective interaction with both suppliers and partners. The system tracks requests, adds notes, records conversations, schedules follow-up contact and helps map out future interaction. CRMs help SMEs streamline their interaction and become better prepared for scaling to a larger size.

CRM is a natural fit for SaaS cloud environments. The biggest advantage relates to how cloud computing removes geographic constraints. Mobile sales forces take full advantage of having access to CRM tools anywhere and at any time. Like all cloud-based SaaS systems, CRMs can be more rapidly deployed. Software updates are automatically managed by the vendor, scalability and elasticity are improved, collaborative capabilities are enhanced, and access is possible from any location with the Internet. Intradepartmental coordination is enabled. Deep integration with social media is possible as are email-based marketing campaigns and other online features. Many CRM software packages offer integration with tools like MailChimp or provide connections to online cloud storage software systems like Dropbox or Box.net.

It is important that an SME thinks about the features they need now and those that may become strategically important. All businesses need access to key features that capture mission critical information. They also must remember "where" their stakeholders reside. Are they social media users? Do they primarily contact the SME via email or phone? Social media tools are cutting edge, but if customers are not in that space, the tracking features may not be helpful. Likewise, SMEs must ensure the CRM software supports apps on the mobile devices used by its team members. The following vendors offer cloud-based CRM software packages[1]:

Salesforce Essentials: Salesforce.com was among the first companies to create a SaaS platform that gripped the collective attention of the business world. Since then, the company has become a gigantic organization with many services, software platforms, and tools. Salesforce provides offerings in all cloud computing areas. One of these, Essentials, is aimed at the SME marketplace. While still considered expensive, it is a version of Salesforce's Sales and Service Cloud product tailored to small business use. It offers smooth workflow processes with advanced features that leverage social media and the web. It is a good package for SMEs that expect to be acquired by a larger organization since data from Essentials can be moved into other Salesforce offerings relatively easily.

Zoho CRM: This is the flagship Zoho product around which their suite of business software operates. It offers workflow automation, marketing campaign development and roll out,

---

1 For more information about CRM tools, see: https://fitsmallbusiness.com/best-crm-for-small-business.

email marketing tools, and a variety of interaction tracking tools. It provides excellent customer management support for a reasonable price and integrates easily with many third-party tools.

Apptivo: This software has a great user interface and is highly customizable for all aspects of customer management. Apptivo focuses on customer support and has a sophisticated degree of security controls. It offers campaign management and workflow management tools with a robust core solution for CRM. Some critics say it lacks integration capability, but these issues have been addressed. This software is part of a broader ERP package.

Freshworks: This is a tool that a small business or startup can use for free. Additional features can be added for a very low cost. This package is user friendly and has a great interface. It is a quick way to jumpstart CRM activities in a small firm. Its primary disadvantage is lack of integration with other tools.

Pipedrive: This is a basic CRM package with a straightforward, easy-to-learn but powerful interface. The company offers an easily configurable package and provides great help and support. Pipedrive runs well from mobile devices and allows its customer information forms to be customized to collect company-relevant information. It offers support for multiple sales initiatives and pipelines. This software specifically is aimed at the SME market.

Considerations for CRM Software Selection

- *Price*: CRM packages offer various pricing approaches. Some require a monthly or annual subscription fee. Others charge upfront and require a maintenance fee. Others require payment for add-ons that provide extra features.
- *Customization Capability*: It is important for CRM packages to permit customization of customer data to be collected. Additional contact fields, custom sales records and other key elements should be considered.
- *Core Feature Set*: Core features should be assessed. These might include contact management, workflow automation, email tracking, campaign organization, sales pipeline management, lead management, analytics, priorities, and so forth.
- *Platform Features:* Does the tool work offline, in mobile environments, and using various computing platforms?
- *Deal and Opportunity Tracking*: The tool should provide a means to track deals and opportunities with clients regarding ongoing sales.
- *Workflow Automation*: Does the tool manage sales processes, follow-ups, and reminders?
- *Reporting Tools*: What reports and features for decision support are included?
- *System Integrations*: Can the tool be integrated with other software currently used in the SME?
- *Customer Reviews*: What do online reviews of the software and customer ratings reveal about the CRM system?
- *Data Management*: How is the data stored and managed? What happens to the data should the SME need to migrate to a different platform, or the vendor goes out of business?

Insightly: This package is a strong contender in the SME marketplace. It provides email tracking and integration with Microsoft Power BI for advanced reporting features. It offers a strong mobile package and focuses on transforming leads into sales. It also has a great user interface with several project management support tools.

Many cloud-based CRM packages exist and are competing for market share. While this is good for SMEs, it does mean some of these companies will be either swallowed up by competitors or shut down. Remember to have a plan in place to migrate data to a new platform should that become necessary.

Project Management/Task Organization

Every business, whether large or small, needs tools to ensure people can cooperatively complete tasks in a time conscious, organized manner. Several new cloud-based SaaS tools are making this possible for SMEs. These project management tools organize teams, track communications, organize documentation, provide reminders, and streamline interaction regardless of location, time, or platform. Several notable tools in this category are Trello, Asana, Basecamp, and Wrike.

Trello: This tool is a cloud-based subscription software package that enables the development of structured to-do lists and project tasks. The user creates boards which can represent projects or initiatives. These boards can be private or shared with another person, a team, or entire organizations. The boards contain lists which can represent a category. The categories can be developed in ways that suit an organization's purpose. For instance, in a project, these might be "To-Do," "Done," or "Priority." Or, the lists might represent specific items that need to be done and are broken into components. The tool permits flexibility and imagination in the creation of an organizational structure. Each list comprises cards. The cards hold details that can represent a task with checklists, details, due dates, resource assignments and so forth. Cards can be moved between lists or boards. They also can be copied, archived, or deleted. An activity panel lets everyone assigned to a project or task quickly see any updates. Trello is very easy to use, intuitive and flexible. Its features permit users to design nearly any type of organizational approach to completing work and communicating. It works best on mobile devices, although the desktop computer application is very robust too. It can be used offline and allowed to sync later when attached to the Internet. One strength is its ability to store images, documents, and other project artifacts for viewing and sharing. For example, it has been used for organizing travel documents including plans, scanned receipts, and reimbursements. It is a great tool if the implementation is carefully designed and rolled out. If not used wisely, it can lead to information overload and confusion. However, proper use makes this one of the best tools invented for task organization. It offers business and enterprise options for the SME with added features like integration with Salesforce, Dropbox, and a variety of other tools.

Asana: This cloud-based project management tool is widely used in IT-related organizations and other companies that engage in complex, multi-person projects. It is an excellent tool for SMEs and offers a free trial with low-cost optional add-ons. It combines features of Trello with traditional project management approaches making it agile and easy-to-use. Teams have workspaces in Asana. Workspaces comprise individual projects which decompose into tasks. Tasks can be documented with comments, notes, or tags. Individual users customize their workspace within an intuitive interface. At its core, Asana facilitates team communication and enables collaboration. It provides ways to follow task and project progress, and share files and other resources with team members. Asana uses a configured email approach with notifications that permit invited people to check, follow, like, or comment on tasks as they are updated. Asana also offers prioritization tools and change management features. Users can customize their personal interface to make the tool easy-to-use.

Asana Features

- Activity feeds
- Automated email messages
- Adding collaborators and attachments to tasks
- Custom calendars
- Custom views
- Task tracking
- Notifications and reminders
- Project and task creation
- Commenting on tasks

- Full mobile support
- Multiple workspaces for team members
- Real-time updates
- Goal, priority, and due date setting
- Ability to view other members' priorities and current tasks when desired
- Project permission customizations
- Task dependency diagrams and Gantt Charts
- Kanban support
- A "My Tasks" priority list

Basecamp: This cloud-based system can be described as a real-time communication tool with built-in project management features. It focuses on making sure everyone is on the same wavelength regarding tasks and projects. This is accomplished with to-do-lists, calendars, due dates, messages, and file-sharing. Basecamp permits teams to manage priorities and track important tasks. Basecamp has a strong mobile application and works well in desktop environments. It is the best project management solution for teams that need to integrate their workflows with those outside the organization. Basecamp has automated connection capability to other tools such as Feed, Dashable, DashStack, enRoute, and others. Basecamp has been in business for nearly two decades and continues to improve and add features. Users organize tasks into to-do lists. Each task has due dates and ranges for completion. Basecamp specifically assigns tasks to employees with features to easily prioritize, reorder tasks, and customize to-do lists. It also includes features for brainstorming and connecting with coworkers using chats and pings. Basecamp offers a document and knowledge repository. In cloud-based environments, this becomes very powerful and impressive.

Wrike: This is a project management SaaS application that can support SMEs or larger enterprises. It offers tiered solutions that make it perfect for an organization that envisions future or rapid growth. The free version supports up to five users and enables a team to add collaborators and use up to 2 GB of storage. The free plan allows users to manage tasks, share files, and monitor activities with a real-time feed feature. To customize and use reporting features, an SME must pay about $10 per user per month. Negotiated enterprise pricing is also available. Unlike Trello, which takes more of a Kanban approach to project management, Wrike promotes a traditional project management approach in a cloud environment meant to enable users to organize and monitor projects and tasks by setting up schedules with priorities and deadlines. Cloud-based features enable information sharing and collaboration using an easy and intuitive interface. Wrike centers on an email collaboration approach and syncs with users' emails. It works to transform emails and replies into visible tasks and syncs these with other users and the platform. A series of dashboards keep these in order with a focus on prioritized tasks in an activity stream. A visual timeline helps keep order as do custom workflows. Budgeting features also can be incorporated. Wrike interfaces with many other tools including Excel, various calendar applications and Google Drive.

Office Software

Office automation software is greatly enhanced by cloud-based technology. In addition to tools to create various documents, spreadsheets, calendars, and presentations, SaaS technologies permit greater levels of sharing and collaboration. Microsoft 365 dominates this area. It is a hosted, cloud-based solution that frees users from needing to install the software on their own in-house servers.

Microsoft 365 is subscription-based and includes Office, Exchange Online, SharePoint Online, Lync Online and Microsoft Office Web Apps. An organization using 365 can administer users from an online portal to set up new company user accounts, control who can access various applications and features, and see current operating statuses of online tools and services.

Microsoft 365 permits administrators to customize features for their organization. The following components are frequently included:

Component	Features/Functions
Office Suite	Word, Excel, PowerPoint, Outlook, OneNote, Publisher, Skype for Business, Access
Exchange Online	email, calendar, tasks
SharePoint Online	web portal for collaboration
Yammer	enterprise social networking
OneDrive for Business	cloud file storage
Power BI	business intelligence

A great deal of information on Microsoft Office 365 exists and its capabilities are well-documented in a variety of online sources. One of the best ways to understand 365 is to compare it to Microsoft's Azure cloud. Both Azure and Microsoft 365 are cloud-based services. The backend components of 365 utilize Windows Azure Active Directory services but to truly understand the two platforms we need to go back to our discussion of SaaS, PaaS, and IaaS. Microsoft 365 essentially is Microsoft's SaaS offering that provides subscriptions to online versions of the popular Office software suite (e.g. Word, PowerPoint,

Access, Excel, Outlook, One Note, and so forth) with SharePoint, Lync and Exchange. Azure, on the other hand, is Microsoft's PaaS and IaaS offerings. While Azure does provide backend services for Office, it is more than just the Active Directory management software for Office. It has many cloud infrastructure and platform options.

Subset of Azure Components

- **Compute:** IaaS permitting virtual machine deployment.
- **Web Apps**: PaaS environment letting developers easily publish and manage websites.
- **Mobile services**: Collects mobile user data and deploys mobile apps.
- **Storage Service**: Provides REST and SDK APIs for storing and accessing data in the cloud.
- **Data Management:** Database, search, data warehouse and other related services.
- **Machine Learning:** Services through Cortana to enhance machine learning.
- **IOT:** Internet of Things infrastructure.
- **Marketplace:** Online applications and services marketplace from Microsoft and partners.

Microsoft 365 is not the only cloud-based office automation software, although it is the market leader by a large margin. Several other packages are worth noting. These include Zoho Office, Google's G Suite, and Apache's Open Office.

Zoho Office: This office suite integrates with all Zoho's software offerings which encompass a broad range of business applications. It has Writer, Sheet, and Show as replacements for Word, Excel, and PowerPoint. Added to these are Projects and Books, their office project management and bookkeeping software. Zoho offers these packages through their cloud system and provides office document storage capabilities. Zoho's software has mobile app equivalents and additional tools for email, system administration, and business analytics. It is a well-rounded package that continues to grow and add functionality. It uses a subscription pricing approach that makes it much less expensive than Microsoft 365.

Google Apps (and G Suite): Google's free online office suite was designed specifically to leverage online collaborative work. It provides applications with functions that mirror many of 365's offerings. Its primary components include Docs, Sheets, and Slides. The applications are less sophisticated than Microsoft's equivalents but are a good option for small businesses or startups wishing to save money since these are free. All Google App documents are saved automatically to Google's cloud storage for each change so that means it is difficult to lose your work. Google also permits users to upload and download their creations in file types common to other applications. For example, it is possible to create something with Google Docs and later move it into Microsoft Word. Google's free cloud storage is limited to 15 GB. For larger organizations, Google offers G Suite. This upgrade provides business email services and higher levels of cloud storage. Several other more expensive tiers also exist with more storage, team capabilities, and a PaaS style platform with an app development environment, smart search, and unlimited cloud storage for organizations with five users or more users. Their top offering, the Enterprise option, includes security, loss prevention and business analytics tools.

Apache OpenOffice: This software has been a Microsoft competitor for many years. It uses an open-source office software suite approach and has components for word processing, spreadsheets, presentations, graphic design, database, and other features. It is free for any use and provides the full set of code which can be customized by sophisticated users. Since it is open source, a community of developers has grown up around it and this means users can report bugs, provide ideas for enhancements, or upgrade the code themselves. Open Office was originally developed and supported by Sun Microsystems. In 2011, the source code was donated to the Apache Foundation by Oracle Corporation which owned Sun. The software is still available for use and deployment but is not recommended as a solution for startups or small businesses unless they possess strong IT expertise.

Other Office Software: Other cloud-based office software applications exist for many purposes. For example, in email automation and marketing, tools such as ActiveCampaign, MailChimp, GetResponse, and ConvertKit exist. These tools enable an organization to automate mailings and then track responses. A related area is survey software. These cloud-based systems permit users to create online surveys, quizzes, polls, or web forms for distribution to stakeholders. Surveys can solicit customer opinions, perform market research and other functions. Survey software can collect information then load it into databases where it can be analyzed. Good survey software permits the creation and customization of questions, has distribution tools, can load data into databases or spreadsheets, tracks responses, sends reminders, and performs analyses. Examples of survey software include: AskNicely, SurveyMonkey, Qualtrics, SurveyGizmo, Wufoo, GetFeedback, Formstack, Zoho Survey, Typeform, Google Surveys, and Google Forms.

Other office software enables an organization to connect with clients via online meetings or video conferencing software. Skype, GoToMeeting, Microsoft Teams, Zoom, and Join.Me are all examples of low cost, cloud-based products that can help in this area.

Paid Features Offered by Many Online Survey Companies

- *Survey logic*: This is the option to create multiple paths through the survey based on answers, random selections, or other characteristics.
- *Export data*: Tools can send survey results to various forms and software interfaces.
- *Target audience*: Some survey companies will provide targeted pools of respondents based on desired characteristics. Qualtrics specializes in this area but it can get expensive.
- *Customizable interface*: This option permits an organization to brand the survey with their logo, corporate colors, or other identifiers.
- *More question formats*: Paid tools often offer a wider range of question formats such as sliders and other graphic-oriented measures.

Data Analytics

Most SMEs cannot afford a full-blown data analytics solution on-premise. This makes using a SaaS application ideal in this area. Many services have emerged in recent years. One of these, Tableau Online is hosted in the cloud. It allows users to upload data, publish dashboards, and share reports via a cloud interface. Tableau specializes in data visualization and the creation of graphic output to represent data. Tableau also offers a free tool called Tableau Public. This tool has most features of Tableau Desktop, except it does not permit users to acquire data from as many sources. And, the results are shared in the cloud. So, this means any mission critical or sensitive data would not be a good fit for the public solution.

Another potentially useful analytics tool is Microsoft Power BI. It comes in both paid and free versions. The free version is limited to smaller datasets which reduces its usefulness. However, the output and visualizations are excellent. Power BI also provides a store for third-party add-on visualization graphics which are usually free or low cost.

Social Media

Depending on the nature of an SME, social media tools can range from very important to no usefulness at all (Wamba and Carter 2016; Hitchen et al. 2017). Many startup plans and new business ideas are focused on social media. For instance, new customer bases, ways of connecting, and marketplaces have emerged due to social media's potential for communicating without geographic and time constraints. New businesses may not need a bricks and mortar store front to ply their trade (McHaney and Sachs 2016; McHaney 2011).

Several basic, cloud-based social media tools can help an SME establish its digital presence. For instance, Hootsuite is a social media listening platform. It has free software that provides services that collect social mentions. This allows a business to maintain a sense of how they are being viewed on social media platforms. Hootsuite Free uses cloud services to enable an SME to manage multiple social networks such as Twitter and Facebook, schedule pre-written social media posts, and interact with stakeholders and followers. Likewise, Hootsuite enables a business to track each social media platform of interest and determine when posts should be released to result in the most views. Hootsuite, like many cloud-based platforms, is highly scalable and if social media becomes more important, a variety of premium plans exist. These provide more capabilities for multiple users within the same business. Hootsuite's competitors include: Buffer, Sprout Social, Canva, TweetDeck, and many more costly offerings including ones from Salesforce (Social Studio) and Twitter (Gnip).

Understandably, SME social media initiatives and strategies often fall into a low priority category. However, not having a social media plan that leverages cloud capabilities, can result in a business without a digital identity. This makes a business seem unreliable or nonexistent to many people. In addition to listening with a social media tool like Hootsuite, several cloud-based tools are worth considering.

- *Facebook*: Every SME should have a presence (e.g. Page) on Facebook. Facebook is a cloud-based giant and is required for establishing an SME's identity. It is too large to ignore and must be included. Facebook also offers Business Tools which enable an SME to see who visits, market products and services, and access analytics.
- *Twitter*: This platform is excellent for connecting with potential customers. It can be monitored to find out what social media buzz exists about an SME or its products and services. It also provides a fast way to get information out to stakeholders. It is easy to have a profile and presence on Twitter.
- *LinkedIn*: This tool generally falls into the domain of business professionals. It provides a place for an SME to establish its credibility and make connections.
- *YouTube*: This social media site provides an excellent place for a SME to post informational videos that can help develop a reputation. Content is king on YouTube and this is where people visit when they need to understand how to do something.

- *Pinterest*: Pinterest is another tool SMEs can consider, particularly if their business focus is on visually interesting products and services. It also is good for organizations with customers who wish to share innovative uses of their offerings. Pinterest works best as a place to showcase uses rather than roll out marketing campaigns.
- *Instagram*: Instagram may not make sense for all SME business plans. It is a good way to connect with stakeholders by sharing photos and other information.
- *Snapchat*: This tool provides a way to connect with customers if an SME focuses on young people. Like Instagram, it offers a visually interesting way to share information.

An SME does not need to develop a presence in all social media tools. Rather, developing a solid business plan that includes use of free and low-cost, cloud-based social media tools can help guide what direction to go in this quickly growing area that offers amazing potential (see Figure 2.8).

What Is a Social Media Strategy?

Social media strategies should be used to show that your SME understands its stakeholders. Knowing your target audience is key to understanding what platforms and content will work best to connect and engage with key individuals. An SME should consider the following:

- *Define a consistent and relevant voice.* Know your audience. Create a document that clearly defines your identity and voice. Include specifics on language use, catch phrases, and sentiment.
- *Listen.* Hear what stakeholders have to say. Set up a system to monitor both what is being said and the sentiment of the messages.
- *Respond.* Social media requires interaction. Manage comments proactively to ensure negative comments are handled professionally. Thank those that provide positive or helpful comments and defuse inflammatory situations rapidly.
- *Post new content regularly.* Develop a schedule for regular postings without annoying stakeholders. Develop a steady, reliable, and regular presence.
- *Balance postings.* Be sure to provide new content, reposted content, and a small amount of promotional material. Specialists in this area refer to the 30/60/10 approach. 30% = new content; 60% = Reposts; and 10% = promotional material.
- *Use scalable, cloud-based platforms that grow as the SME grows.* Hootsuite and other tools have free packages easily upgraded as needs change.

Purchasing and Procurement

Purchasing, procurement, and vendor management benefit from cloud-technologies. These areas help SMEs find, buy, and manage supplies and services required to conduct business. A wide range of software exists in this area. Functionality may be built into accounting or inventory management software systems or found as standalone packages meant to run in SaaS environments. Reviewing several packages helps describe the functionality of purchasing/procurement software.

Procurify: This cloud-based, purchasing solution offers features to track and control inventory, and manage costs. This software package enables teams within an organization to make purchases and then consolidates all organizational purchasing information to help find volume discounts and the best vendors. Since it is cloud-based, Procurify has a real-time approach to purchasing. It lets users create purchase requisitions then track the goods or services as they are acquired. It has features to allow ratings of the transactions. The interface provides a "shopping cart" type experience and permits an in-app messaging and notification system. Powerful analytics and reporting systems provide information about purchases and expenditures. These can be integrated with other cloud-based software, particularly accounting or ERP systems. It also offers mobile versions of the software.

Snapfulfil: This cloud-based software solution focuses on procurement and purchasing from a warehouse management perspective. It is meant for use with Business-to-Business (B2B) as well as Business-to-Customer (B2C) industries. One of Snapfulfil's strengths is its cloud-based approaches for information storage that ties tightly to physical space management within an organization. For instance, it has a space utilization module to help design optimal floor layouts for storage. It also includes route optimization features for delivery planning. Snapfulfil provides full labelling and parcel tracking features with a focus on supply chain management.

Infoplus: This cloud-based software provides an inventory management solution with specialized features for industries such as consumer goods, food and beverage, and retail. Infoplus offers several physical inventory-related features including demand forecasting, inventory optimization, kitting, vendor-managed inventory, and lot sizing tools. It also integrates closely with ordering channels such as Amazon.com. The software was developed to permit use of mobile devices by people working in warehouse situations.

Help Desk and Service Software

Customer service software features often are integrated into CRM solutions. However, specific cloud-based software exists to help with this area, particularly for IT departments that support many users. Many SMEs can save time and resources using software developed for tracking and managing customer issues. In general, help desk software enables SMEs to resolve customer concerns or problems by managing tickets generated to track problems from receipt to resolution. These packages also provide tools to resolve issues, to track and ensure solutions are not reinvented each time a similar issue arises, and to catalog problems so preventative solutions can be developed. Many concepts and approaches used in these areas were developed in conjunction with software support industries. A primary function for helpdesk applications is to ensure customer support is consolidated and not run from the email accounts of several people where communication can get lost or overlooked. Instead, having a centralized, virtual location for requests and tracking means support is unified, and multiple people can collaborate to identify, resolve, document and prevent issues. Several good SaaS software systems exist in this area:

LiveAgent: This is a leading software package that offers many tools and features. It is expensive, comparatively, but supports VOIP tools and other customer interaction channels. It has unique features like self-service tools and help items, and multiple service level capabilities so particular customers can be given more help or priority service.

Zendesk: This highly popular package focuses on communication and keeping customers in the loop. This is an excellent choice for small companies or startups seeking to keep costs low. It is easily scalable and provides an excellent ticket tracking and notification system. It is also a good choice for companies with their support staff located remotely.

Freshdesk: This software can either be standalone or used with other tools from the same company. It has a free basic package that can be quickly scaled upwards. It has an easy data import feature and ensures data belongs to the company using the software, not Freshdesk. This is helpful in many ways. For instance, help operations can be rapidly moved to another software system if desired, and in sensitive situations, data can be backed up or removed quickly. This also means outside analytics and other tools can be used

Desk: This SaaS system is part of the Salesforce software family. It is widely considered to be the most powerful and fully featured helpdesk package. It uses a shared inbox for the help desk staff and includes capabilities for automating help functions. It also has a powerful mobile interface. An interesting feature of this software is its "customer health" metric which helps identify frustrated customers needing extra attention.

Key Help Desk Software Features

- *Ticket management*: Tickets are essentially a customer request for help. Good help desk software filters and categorizes tickets so they can be assigned to team members, prioritized, and tied to customer history. Tickets can be viewed by all relevant people and ensure everyone is on the same page regarding resolution and customer communication.
- *Support for multiple communication channels*: Customers may contact the support team via email, phone, messaging, posts, chats and so forth. Good help desk software permits all that make sense and helps to ensure consistent documentation no matter what channel is used.
- *Customer help and FAQ area*: Many issues can be quickly resolved with standard help documents. Videos and other customer education options can be helpful too.
- *Collaboration*: Enabling a team to work on tickets through SaaS features is a helpful addition to most help desk systems.
- *Reporting*: A series of reports and statistics regarding resolution times and other key metrics is important.
- *Deadline tracking*: Having notifications and reminders about overdue tickets is a useful feature in these software systems.
- *Mobile interface*: Having a good mobile interface gives both customers and help team members added capabilities.

Enterprise Resource Planning (ERP)

Enterprise Resource Planning (ERP) software essentially refers to a suite of business tools meant to automate all basic business functions. Generally, ERP software integrates applications using a consolidated database and ties all business functions to an accounting system. Typical items integrated into an ERP system range from sales to order fulfillment. Often included are human resources, CRM, invoicing, supply chain, inventory control, finance, project management, procurement, accounting, and many other business functions. A complete ERP package removes the need for many other software packages.

ERP software traditionally has been the domain of large corporations with complex operations. Recently, many new products using SaaS, have become available to SMEs. This means, that many business functions once limited to only large businesses are now feasible for smaller companies.

ERP systems have many advantages to offer an organization. For instance, software is standardized across a company. All data is stored in a single location which reduces the likelihood of redundant, mismatched data. ERP systems remove the need to integrate separate tools and data features across different platforms. This reduces the need for support teams and lowers software subscription costs.

ERP systems often provide better insights since organizational data is consolidated. Many ERP packages either include business analytics tools or have features to easily move data into visualization tools like Power BI or Tableau. ERPs offer custom dashboards that display important metrics to the people that need to see them, and permit customization of interfaces for various work roles to ensure the correct data is in front of the people that use it for their jobs. For example, salary information will be accessible to HR managers but not to a purchasing agent. Business rules can be configured to match the organization's operations. And, most ERPs include best practice workflows that can make business tasks easier to complete.

SMEs often use a subset of ERP features and as they grow, scale up into a broader set of functions. Many ERP vendors offer subscription prices that allow adding or removing functional areas. The ERP marketplace for small organizations has many choices. For mid-sized organizations, Microsoft Dynamics, NetSuite ERP, and SAGE are among the leaders. For smaller businesses, Apptivo and Work[etc] are good choices in a crowded, competitive market. Figure 2.9 provides a depiction of ERP software functionality.

Apptivo: The Apptivo CRM package discussed earlier in this chapter offers ERP functionality specifically for SMEs. It offers support for marketing, purchasing, supply chain management, and limited accounting functions that include invoicing and expense tracking/reporting. Although it offers a free, three-user version, a paid tier must be used to integrate the package with software like QuickBooks to obtain a full set of accounting features. Apptivo is a good example of ERP functionality available at a low cost to a small business which can be upgraded when needed.

Work[etc]: This is an ERP for SMEs from a CRM and project management perspective. It offers features for marketing, sales, customer support, help desk, operations management, and limited financial tracking. It easily integrates with many accounting software systems for real-time reporting and dashboard use.

Corporate Managers and Users

Cloud services are very important to large corporations and are a primary focus of Chief Information Officers (CIOs), IT managers, and many users. As with SMEs, large organizations use SaaS platforms to reduce the cost of software ownership and use. The same SaaS platforms and storage options used by individual users and SMEs often are scaled for use by those working in the largest businesses. Although initial resistance existed for a move to cloud computing, most Fortune 500 companies have incorporated various levels of cloud computing into their operations.

In general, corporate managers rely on cloud computing to reduce volatility and enhance elasticity. This means traditional cycles of buying hardware and software in waves have been reduced or eliminated. Applications in the cloud are easier to configure and maintain, and changes often are transparent to end-users. Just like SME users, corporate users of SaaS services find applications quickly and efficiently scale to meet changing capacity needs. This translates to stable, predictable savings and reduced budgets. Nearly all corporate employees use some form of cloud computing. Many of these uses will be discussed in more detail later in this book. A general overview of these follows.

Organizational Users of Cloud Computing

Nearly everyone in an organization is a cloud computing user. In addition to those using SaaS platforms and ERP systems, most users fall into several categories. Among these are PaaS, IaaS, storage, backup and recovery, and data analytics users.

PaaS Users

Cloud computing using PaaS has transformed DevOps in organizational settings. *DevOps* (which stands for development and operations) focuses on enterprise software configuration and development in a way that tightly links it to IT operations in the form of an agile relationship. DevOps relies on close relationships between development and operations and advocates enhanced communication and collaboration between teams representing these business functions. PaaS provides a development and deployment cloud-based environment. This enables IT users to communicate, collaborate and work together in a concerted, agile manner. PaaS users have access to resources from a cloud server provider supplying a platform like Amazon Web Services or Microsoft Azure. Resources are purchased as needed and accessed over the Internet.

DevOps has become the acronym that represents a combination of philosophies, practices, and cloud-based tools used to enhance an organization's capability to design, build, configure and rapidly deliver applications and services. DevOps seeks to improve products at an accelerated pace. It is an evolutionary step past traditional systems development and IT infrastructure management. DevOps offers best practices to achieve its goals. Among these are:

- *Continuous Integration*: Combine development and operations in a constructive and facilitative manner.
- *Continuous Delivery*: Updates are ongoing. Problems are fixed and enhancements are offered transparently to users.
- *Microservices*: Applications are a loosely coupled collection of services which can be combined to enhance business operations (e.g. Lego blocks).
- *Infrastructure as Code (IaC)*: Data center provisioning and management is conducted with rule-based script files rather than physical hardware configuration.
- *Monitoring and Logging*: User activity is monitored to identify problems or potential areas for optimization. Logs are created and analyzed.

> • *Communication and Collaboration*: DevOps focuses on enhanced communication between developers and IT infrastructure specialists.

IaaS Users

IT managers, particularly those on the operations side of DevOps use IaaS to leverage resources in many areas. For instance, web servers, storage, backup and recovery, high performance computing, and number of other areas. In general, IaaS removes the need for capital investment and helps reduce recurring hardware costs. Onsite data centers, physical servers, and other hardware items are virtualized, and replaced with subscription services.

File Storage and Backup Users

Cloud computing started with the concept of off-site storage and this remains one of the most widely used and economical cloud applications. The cloud facilitates access, retrieval, safe backup, and real-time usage of data. Cloud storage makes sense from a financial sense as well. Users only pay the storage costs for what they consume. From a backup perspective, cloud storage makes sense. For instance, automatic scheduling algorithms ensure data is safely stored in remote, secure locations. Stored files are available without time or geographic constraints. And, scalable storage capacity exists.

Disaster Recovery Users

Business data can be safely stored in the cloud. Even if a business is physically compromised, its data is safe in multiple, cloud-based locations. Disaster recovery in cloud environments is more effective and economical.

Big Data Analytics Users

Cloud computing enables data to be easily accessed and analyzed using data analytics tools. Cloud-based environments have enabled businesses to collect, manage, and analyze vast quantities of structured and unstructured data. Data from multiple sources is combined in the cloud to provide new insights. This is particularly true in the age of social media.

Summary

Cloud computing is pervasive and ubiquitous. Users come from businesses and organizations of all sizes, shapes, and forms. Individual users also find the cloud appealing and have incorporated SaaS into their lives using social media, personal software subscriptions, and online storage of media and personal artifacts. This chapter has provided an overview of cloud use from individual, SME, and corporate perspectives together with examples of the software, services, platforms, and infrastructure each uses. Particularly, we focused on individuals and SMEs because corporate users form the remaining book chapters' focus.

References

Cheng, C.C.J. and Shiu, E.C. (2019). How to enhance SMEs customer involvement using social media: the role of social CRM. *International Small Business Journal* 37 (1): 22–42.

Hitchen, E.L. et al. (2017). Social media: open innovation in SMEs finds new support. *Journal of Business Strategy* 38 (3): 21–29.

McHaney, R.W. (2011). *The New Digital Shoreline: How Web 2.0 and Millennials Are Revolutionizing Higher Education*. Herndon, VA: Stylus Publishing.

McHaney, R.W. and Sachs, D. (2016). *Web 2.0 and Social Media: Business in a Connected World*, 3e. Copenhagen: Ventus/Bookboon.com. [Online]. Available at: http://bookboon .com/en/web-2-0-and-social-media-for-business-ebook.

Wamba, S.F. and Carter, L. (2016). Social media tools adoption and use by SMEs: an empirical study. In: *Social Media and Networking: Concepts, Methodologies, Tools, and Applications*, 791–806. IGI Global.

Further Reading

Al-Johani, A.A. and Youssef, A.E. (2013). A framework for ERP systems in SME based on cloud computing technology. *International Journal on Cloud Computing: Services and Architecture* 3 (3): 1–14.

Gracia-Tinedo, R. et al. (2016). Understanding data sharing in private personal clouds. In: *2016 IEEE 9th International Conference on Cloud Computing (CLOUD)*, 392–399.

Hohemberger, R., Rossi, F.D., Konzen, M.P. et al. (2020). Towards elasticity for SME cloud services. In: *2020 International Conference on Information Networking (ICOIN)*, 541–546. IEEE.

Ross, P.K. and Blumenstein, M. (2015). Cloud computing as a facilitator of SME entrepreneurship. *Technology Analysis & Strategic Management* 27 (1): 87–101.

Seay, C., Washington, M., and Watson, R.J. (2016). Personal applications of clouds. In: *Encyclopedia of Cloud Computing*, 517–523.

Zhang, G. (2017). Exploring vendors' capabilities for cloud service development and delivery. Doctoral dissertation. Loughborough University.

Zhang, G. and Ravishankar, M.N. (2019). Exploring vendor capabilities in the cloud environment: a case study of Alibaba cloud computing. *Information & Management* 56 (3): 343–355.

Website Resources

Accounting Software

QuickBooks: https://quickbooks.intuit.com
Sage One: https://www.sageone.com
Xero: http://www.xero.com
Zoho Books: https://www.zoho.com/books

CRM Software

Apptivo: http://www.aptivo.com
Freshworks: https://www.freshworks.com/freshsales-crm
Insightly: https://www.insightly.com
Pipedrive: https://www.pipedrive.com
Salesforce Essentials: https://essentials.salesforce.com
Zoho CRM: https://www.zoho.com/crm

Data Analytics

Microsoft Power BI: https://powerbi.microsoft.com
Tableau: https://www.tableau.com

ERP for SMEs

Apptivo: https://www.apptivo.com
Work[etc]: https://www.worketc.com

Help Desk

Desk: https://www.salesforce.com/solutions/small-business-solutions/help-desk-software
Freshdesk: https://freshdesk.com
LiveAgent: https://www.liveagent.com
Zendesk: https://www.zendesk.com

HR Software

BambooHR: https://www.bamboohr.com
EffortlessHR: https://www.effortlesshr.com
Namely: https://www.namely.com
SAGE HR: https://www.sage.com/en-us/products/sage-hrms
SuccessFactors: https://www.sap.com/cmp/ppc/na-human-experience-mgmt/index.html
TidyForm: https://www.tidyform.com
Zenefits: https://www.zenefits.com

Office Software

Apache OpenOffice: https://www.openoffice.org
Google Apps (and G Suite): https://gsuite.google.com
Microsoft 365: https://www.microsoft.com/en-us/microsoft-365
Zoho Office: https://www.zoho.com/docs/office-suite.html

Project Management Tools

Asana: https://asana.com
Basecamp: https://basecamp.com
Trello: https://trello.com
Wrike: https://www.wrike.com

Purchasing and Procurement

Infoplus: https://www.openoffice.org
Procurify: https://www.procurify.com
Snapfulfil: https://www.snapfulfil.com

Social Media

Facebook: https://Facebook.com
Instagram: https://Instagram.com
LinkedIn: https://LinkedIn.com
Pinterest: https://Pinterest.com
Snapchat: https://Snapchat.com
Twitter: https://Twitter.com
YouTube: https://YouTube.com

3

What Is Virtualization?

Cloud computing is complex both from conceptual and technological perspectives. This chapter provides insights into several key concepts and enabling technologies that make cloud computing appeal to a wide range of users. We start by exploring virtualization in this chapter. Cloud computing relies heavily on the concept of virtualization (Portnoy 2016). It is important to view virtualization as an enabling technology for cloud computing, especially for IaaS. It is not an interchangeable term for cloud computing but rather part of the system used to enable scalability, elasticity, management of resources, and deployment of solutions.

Even if infrastructure was completely virtualized, other considerations would require attention. For instance, process automation and tools to manage applications on the virtualized machines could be lacking. So, for now, think of virtualization as a key, enabling technology that makes cloud computing possible.

When most users consider the cloud, the first thing that comes to mind is large pools of shared data. These same users may not realize they share infrastructure through virtualized machines. This remains hidden to most people (the non-IT ones in particular) with desktop computing applications which can be used from any PC in their organization or home. In many cloud applications, this unseen technology virtualization becomes the primary benefit. Users receive the current version of standard applications from a host who maintains and updates the software in exchange for a subscription fee.

Put differently, virtualization is a technique used by IT specialists to enable groups of users to share physical instances of a technology resource (Barrett et al. 2010). These users do not necessarily have to be from the same organizations. IT specialists assign logical names to physical resources and provide logical pointers enabling it to be used when needed. So, virtualization creates a logical (or virtual) version of a resource, such as a storage device, server, computing desktop, operating system, or network resources that can be accessed by users transparently and without added effort. We will examine hardware virtualization, operating system virtualization, containerization, and storage virtualization in the chapter.

Cloud Technologies: An Overview of Cloud Computing Technologies for Managers, First Edition. Roger McHaney.
© 2021 John Wiley & Sons Ltd. Published 2021 by John Wiley & Sons Ltd.
Companion website: www.wiley.com/go/mchaney/cloudtechnologies

Hardware Virtualization

Using virtualization techniques and software to create multiple virtual machines on an existing, physical computing platform is called *hardware virtualization.* Virtualization logically separates the created environment from the underlying hardware. The physical hardware component is known as the host (see Figure 3.1).

On a non-virtual, physical computer, operating system software (e.g. Windows or Linux) is installed and has access to the computer's actual hardware and connected components. This means the CPU, memory, disks, other storage, and so forth are controlled directly by the OS. For an isolated, individual user, this is fine but in organizational settings, ensuring multiple computers are all maintained can become time consuming and difficult. In the past, whenever new software versions needed to be rolled out or updates completed, a great deal of human interaction was required. Virtualization helps automate these tasks and reduce human labor.

Software, known as the virtual machine manager (VMM) or hypervisor, is installed on a physical machine (a server or servers with greater capacity) and is used to create virtual machines. The virtual machines are controlled and monitored, and utilization of the physical hardware (e.g. processor, memory, and other hardware resources) is tracked. Once the virtual machines have been provisioned into instances on the physical hardware, they can be loaded with various operating systems and software applications tailored to user needs. This end-user software relies on virtual computing resource representations (e.g. virtual processors, virtual disks, and so forth). Popular hypervisor software packages are Microsoft's Hyper-V and VMware's vSphere.

Virtual machines (VMs) are logically independent and isolated so each has its own custom set of software, including operating systems if desired. Even though host hardware can run multiple VMs simultaneously, each is logically independent. The idea is to keep one VM's crash or virus infection from impacting others.

Hypervisors

A hypervisor, or VMM, is used to create, run, and monitor virtual machines. A hypervisor most commonly is deployed as software but can be firmware or hardware too. The hypervisor runs on a host machine that will house the virtual machines, sometimes called guest machines. Primary hypervisor functions are to ensure guest machines have a virtual hardware platform and then manage requests from guest machines' operating systems in interactions with the virtual hardware platform. Hypervisors permit multiple instances of operating systems to operate on isolated VMs simultaneously.

The term *bare-metal virtualization* often is used to describe the installation of a hypervisor as an operating system. In this instance, the hypervisor or VMM is installed directly on computing hardware. For example, Microsoft's native hypervisor software, Hyper-V, has been included with Windows Server since the 2008 edition. It can be implemented in two ways. First, it can be a component of Windows server capable of creating, running, and monitoring virtual machines. Second, it can be used as a self-contained Hyper-V Server OS. The second scenario is a bare-metal installation. When used as a bare-metal virtualization, other OSes and applications are installed later. Figures 3.2 and 3.3 illustrate.

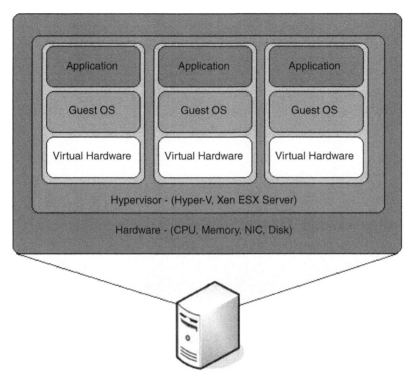

Figure 3.1 Hardware virtualization creates multiple logical machines on shared hardware. Source: Kwesterh, Local diagram of hardware virtualization, 2010.

Hypervisors integrated into an operating system are implemented through processor command set extensions. These extensions were developed by chip manufacturers to enhance speed performance in microchips likely to be used in servers hosting VMs. Example extensions include Intel Virtualization Technology (Intel-VT), AMD Virtualization (AMD-V), and VIA virtualization (VIA VT). Nearly all modern microchips include virtualization extensions. However, some machines have the capability turned off by default, but can be enabled with a change to BIOS settings (see Figure 3.4).

Host Virtualization: Some IT shops used a technique known as Host Virtualization. This was an alternative to the bare-metal approach. In this instance, the OS is installed first then standalone hypervisor software was installed. VMs were created on this platform. This approach has fallen into disuse due to integration of many OSs with hypervisor functionality. However, in recent years container-based virtualization, using the same ideas, has gained in popularity.

Types of Hardware Virtualization

Hardware virtualization is implemented in a variety of ways. Among the most popular approaches are full virtualization, paravirtualization, and hardware-assisted virtualization.

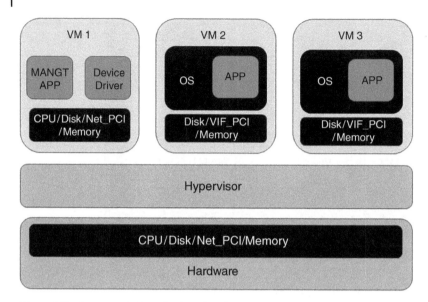

Figure 3.2 Bare-metal virtualization showing hypervisor running three virtual machines. Source: Adapted from Radoslaw Korzeniewski, XEN Hypervisor schema, 2007.

Full Virtualization: This virtualization approach seeks to emulate the full set of hardware needed to run a VM in a hosted environment. A fully virtualized instance runs on top of a guest OS. The guest OS runs on the hypervisor which runs on the host's operating system and hardware platform. This form of virtualization is meant to be indistinguishable (from an application standpoint) from an actual hardware platform (see Figure 3.5).

Paravirtualization: This virtualization approach requires the guest operating systems to be modified to work on the virtual machines running on the host. Hardware is not exactly emulated. To counter this, the VMM uses an application program interface (API) approach. It modifies the guest machines' operating systems and replaces the existing code with instructions to enable calls to the VMM's APIs. The hypervisor recompiles the modified OSs. The calls between guest OSs and the hypervisor are referred to as *hypercalls*. This can become very complicated. In general, paravirtualization is used to improve overall host performance by decreasing the overall number of VMM calls. Xen is a virtualization product often used in paravirtualization approaches.

Hardware-assisted virtualization: This technique uses the host computer's existing hardware to build a VM. This technique was first used in the early 1970s by IBM with its System/370 mainframes. Due to how slow software ran in those times, engineers developed a hardware-based solution to speed things up. Although the original hardware-assisted virtualization techniques are obsolete, the ideas influenced development of today's microprocessors and their extended command sets that provide virtualization options. Today's processors have built-in commands to help create VMs and this improves performance of host computers and enables a server to maintain more VMs with an efficient overall performance. This approach to hardware-assisted virtualization currently is in widespread use.

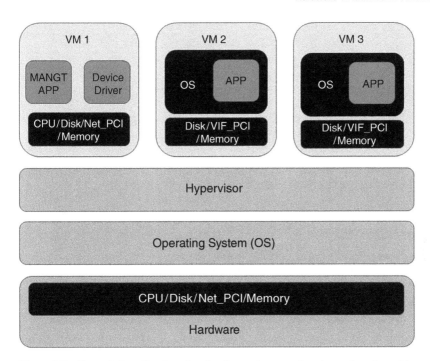

Figure 3.3 Hosted virtualization showing hypervisor running three virtual machines.

Graphics Virtualization Technology (Intel GVT-d, GVT-g and GVT-s): Another popular hardware virtualization relates to graphics technology. Intel developed an integrated Graphics Processing Unit (GPU) called the Iris Pro. It was designed to be assigned to a specific VM machine or shared between multiple VMs. This hardware virtualization revolutionized video game speeds and made it possible for video related processing to be moved off the main CPU.

Hardware Virtualization Vendors and Products

Many vendors play important roles in the hardware virtualization marketplace. This is a difficult field to enter and requires highly specialized, technical products. Among the leading products in hardware virtualization are VMware ESXi, Microsoft Hyper-V, and Xen.

VMware ESXi: This hypervisor continues VMware's traditional leadership role in hardware virtualization. Its name came from Elastic Sky X (e.g. ESX), the original version of the software. Later, after an upgrade which removed a slow service console component, the "i" was added to represent the word "integrated." ESXi is installed on a host computer and assumes complete control over the machine's hardware resources. It does not need an OS to be installed on the host and includes its own kernel (see Figure 3.6).

Microsoft Hyper-V: This hypervisor takes advantage of Microsoft's knowledge of Windows operating systems using x86 architecture microchips. Hyper-V creates partitions and isolates VMs, so each guest OS executes a partition. The partitions operate in a parent and

Figure 3.4 Virtualization disabled in an older computer's BIOS. Source: By DELL-BIOS(VT) | by ryuuji.y [Some Rights Reserved] http://www.flickr.com/photos/ryuuji_y/.

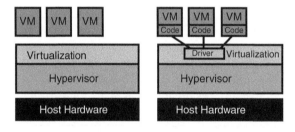

Figure 3.5 Full virtualization (on left) versus paravirtualization (on right).

child configuration. Parent partitions are given direct access to all system hardware. Each child partition only has a virtual view of system resources. A parent creates a child using a hypercall API. Hyper-V is used in 64-bit versions of Windows 8 and above.

Xen: This hypervisor was developed to be an open source product as part of the Linux kernel. Its development is overseen by the Linux Foundation. Xen is not used on all Linux distributions despite its full virtualization, paravirtualization and hardware-assisted virtualization capabilities.

Hardware Virtualization Benefits

Although independence of VM instances is a good benefit, organizations often utilize virtualization to enhance resource management and improve hardware utilization. This is particularly true in cloud environments. Rather than buy multiple servers, an organization can host and manage multiple instances on one piece of hardware. Taking this concept to the cloud and using it as a key component of IaaS means the utilization and benefits are extended to multiple organizations or individuals, and management can be handled by a team of experts.

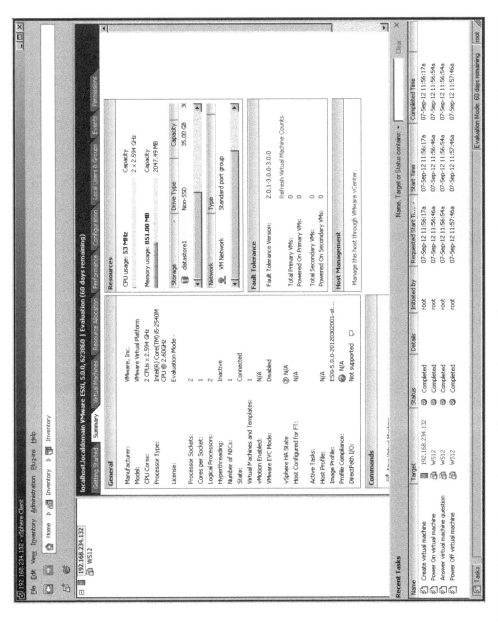

Figure 3.6 VMware ESXi evaluation version screenshot. Source: "vmware player vmware esxi vm vmware vsphere client" by osde8info. licensed under CC BY 2.0. Found at: https://www.flickr.com/photos/osde-info/7948643886.

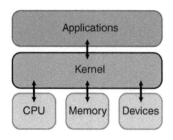

Figure 3.7 Kernel's role in a CPU.

Kernel: Centerpiece of a computer's operating system. This program controls all other programs. The kernel provides services that enable other programs to request various hardware items central to the computer. It also sends requests to device drivers that control attached hardware. The kernel regulates CPU use and enables multiple programs to use it simultaneously (e.g. multitasking). In everyday language, the kernel is the center of the computer's brain! See Figure 3.7.

Operating System Virtualization

Operating system virtualization is a term often used to describe a technique where VM software is installed on the host computer's operating system. This contrasts with bare-metal virtualization. This technique has been used to run older software that requires outdated operating systems to run. For instance, a virtual version of the DOS operating system can be installed on a windows-based platform to enable a user to play an old video game. As an example, the program Virtual DOS machine (VDM), allows a virtual 16-bit/32-bit version of DOS to run software using another host Windows-based platform. Another similar example is DOSBox. DOSBox is software that emulates a full CPU that can run DOS programs. It relies on virtualization capabilities built into the x86 family of processors (see Figure 3.8). Both VDM and DOSBox rely on having a host computer's operating system interact with physical hardware to enable virtual operating systems.

Operating system virtualization can permit a variety of virtual operating systems to share virtualized hardware resources. Besides the DOS emulators already mentioned, Linux, a variety of Windows versions, and MAC OS versions can run on a single physical x86-based host. This contrasts with operating-system-level virtualization or containerization which we will cover in the next section. The terminology is confusing, but a big difference exists. In current development environments, operating system virtualization permits testing software applications on different operating system platforms without needing to have physical hardware and installed operating systems of each sort. Sometimes this use is known as application virtualization.

Operating-System-Level Virtualization (Containerization)

This form of virtualization, though like operating system virtualization, is quite different. It uses a host operating system feature where the kernel permits creation of isolated, user-space instances called containers. Sometimes these containers are called partitions or

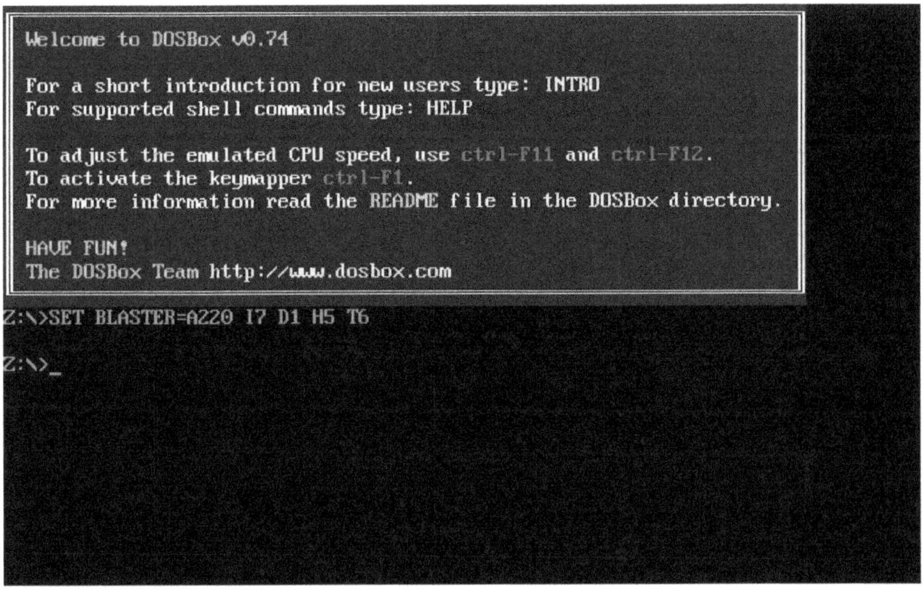

```
Welcome to DOSBox v0.74

For a short introduction for new users type: INTRO
For supported shell commands type: HELP

To adjust the emulated CPU speed, use ctrl-F11 and ctrl-F12.
To activate the keymapper ctrl-F1.
For more information read the README file in the DOSBox directory.

HAVE FUN!
The DOSBox Team http://www.dosbox.com

Z:\>SET BLASTER=A220 I7 D1 H5 T6

Z:\>_
```

Figure 3.8 DOSBox running on a host computer with a Windows operating system. Source: By SF007 at en.wikipedia. Later version(s) by Tornedo500, McLoaf at en.wikipedia. [GPL (http://www .gnu.org/licenses/gpl.html)], via Wikimedia Commons.

jails. Containers provide space for applications and these applications do not experience any difference from running solo on a physical machine – at least that is the goal and concept. The applications have access to all resources such as disk drives, files, network connections, and hardware items. But, are limited to those the container has permission to use. The kernel controls all allocations.

The idea of containers was developed to enable applications to run in efficient isolation on a shared operating system kernel without needing the deployment of VMs. Figure 3.9 illustrates. The diagram refers to "containerization software." This usually is a commercial software package such as Docker or Linux Containers (LXC) that enables the operating-system-level-virtualization to occur. Notice that no hypervisor nor guest operating system exists in this configuration. A primary reason for using this approach is that the virtualization requires little overhead. Containerization software generally includes a high-level API to implement lightweight containers that run isolated applications.

Containerization Software

Several commercially available containerization software systems are available. Among these are Docker and LXC.

Docker: This software was first released in 2013 by Docker, Inc. It has met with wide acceptance and is used extensively in environments needing to test web servers and web applications. For example, a software specialist using Docker implements a container that runs a web server and application. A second container runs a database server used by the application. The two containers are isolated and cannot be seen by one another. That way, the interface between the application and database can be tested in a variety of guest

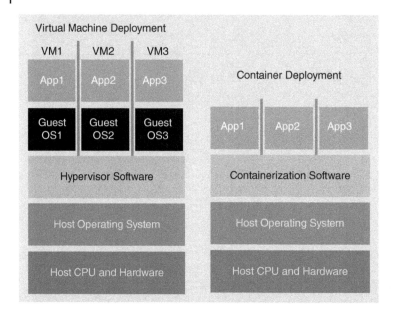

Figure 3.9 Containerization differs from standard virtual machine deployment.

OS environments scenarios. However, all containers are run by a single OS kernel. *Images* are used to specify the containers' contents. Tested images can be ported to other virtual or real machines. Containers are flexible and can be run on public clouds, bare-metal installations, private clouds, or actual physical machines. Docker has been integrated into several IaaS services including AWS, Google Cloud Platform, Microsoft Azure, and others. Microsoft integrated the Docker engine into its Windows Server 2016 platform to enable Docker to be used natively on Windows 10+.

Docker's Components

Daemon: Docker uses a persistent process to manage user-defined containers and container objects. This process, the Docker daemon, is called *dockerd*. It listens for requests sent to the Docker API.

Client Program: Docker's client program, also called "docker," is a line interface that permits users to directly interact with the daemon.

Objects: Objects are used to build applications within Docker. A wide range of objects can be used. The primary object classes include containers, images, and services.

- *Containers*: Docker containers are a standardized, encapsulated environments used to run applications. A container instance is managed using the Docker API.
- *Images*: These are read-only templates used to build the contents within a container. Images are used to deploy applications after testing and debugging usually in multiple locations.
- *Services*: These permit scaling containers across multiple Docker daemons. A swarm of cooperating daemons are created, and these daemons communicate through Docker's API.

> *Registries*: Docker Registries are Docker's repository for retaining images. Clients connect to registries to allow images to be downloaded (or uploaded). Docker offers both public and private registries.
> *Hub*: Docker Hub is the default public registry.
> *Cloud*: Docker Cloud is a native solution used to deploy, monitor, and manage Dockerized applications.

Docker offers two tools to help manage and deploy applications: Docker Compose and Docker Swarm. Docker Compose allows IT specialists to define and run multi-container applications. It helps with application configuration, creation, and start-up of the containers. It includes utilities that enable users to simultaneously run commands on multiple containers. Docker Swarm is a clustering tool that enables users to transform a group of Docker daemons into a single, virtual Docker engine.

Linux Containers (LXC): While Docker is considered an application container, intended to support development, testing, and deployment of applications, LXC is a system container. System containers occupy the space between VMs and application containers. A system container can run an OS but does not emulate the underlying hardware specific to an OS. So, system containers operate like application containers but allow the developers to install different library versions, development languages and databases. This can be helpful if multiple versions of an operating system (or database, et cetera) need to be tested on the same hardware platform.

Hyper-V and Windows Server containers: Microsoft also offers container software. Hyper-V containers use a small footprint VM to provide application isolation, making it like LXC's system container approach. Application isolation in Windows Server containers uses namespaces, resource control, and other tools provided by the server software.

Kubernetes: Although perhaps not technically containerization software, Google's former product, Kubernetes (now open source), organizes containers into pods located on nodes, which is their term for hosting resources. Kubernetes is used for many container functions and helps automate, deploy, scale, maintain, and operate application containers. Kubernetes is considered an orchestration tool like Docker Swarm.

Containers versus Virtual Machines

Although they perform similar tasks, containers and VMs are different. It is easy to confuse the terms operating system virtualization and operating-system-level virtualization (containers). Containers take less memory than an equivalent VM. VMs each emulate a hardware system meaning multiple OSs can run in independent environments on the same physical machine. Hypervisor software emulates the hardware required from pooled (particularly in cloud environments) CPUs, memory, storage, and network resources. The hypervisor ensures the resources can be shared by multiple VMs. Therefore, VMs take more memory since each one requires a guest OS. Containers, on the other hand, share the hosts' OSs. VMs may be larger than an equivalent container and take more time to run and create. However, they have the added flexibility of permitting multiple OSs from various vendors. They also can run multiple versions of OSs, including older ones.

Deciding to use a VM or container may depend on development team goals. For instance, a Web application may be used on a variety of client machines on various browsers being run on various OSs. In this case, a VM makes more sense. Whereas, if a homogenous environment was being used for an in-house software project, containers might be better. These are very simplistic examples meant to illustrate the concepts of containers and VMs. Specialists spend a great deal of time working in DevOps environments that require innovative approaches to software development and deployment, and their judgment often determines the approach to use.

Benefits of Containers

- Can be more efficient than VMs since they share the host OS kernel.
- More portable due to encapsulation of the application operating code.
- No guest OS environment variables or library dependencies need to be managed.
- More efficient use of memory, CPU, and storage.
- More containers can run on same infrastructure as VMs.
- Consistent through entire application lifecycle (particularly in agile development).
- Better for distributed applications.

Disadvantages of Containers

- Lack of isolation from the host OS.
- Security threats may be greater than for VMs.
- Lack of OS flexibility. Each container generally uses same OS as host.
- More difficult to monitor activity on containers since many more can simultaneously exist.
- Large number of vendors and orchestrators in this field make the technology complicated.

Container Cloud Practices

Containers are well suited for cloud use, particularly with microservices development. For instance, a microservice that logs system use can be composed of containerized services. Each service can run on the same OS but remain isolated. As user demand ramps up, the services can be scaled up. This is perfectly suited for use in cloud environments where elastic, on-demand processing is desired.

Microservices

Microservice architecture describes a software development approach where an application is broken into smaller, specialized parts that communicate across REST-based interfaces or APIs. In general, microservices manage their own data, authenticate and maintain logs. Microservices are ideal for container environments.

Containers as a Service (CaaS)

This is a cloud-based service model where containers and related applications can be uploaded, organized, and managed to facilitate use of this technology. Currently, all major PaaS cloud vendors offer CaaS services. For example, Amazon has its Elastic Container Service (ECS) and AWS, Fargate. Google has Google Kubernetes Engine (GKE). Microsoft offers its Azure Container Instances (ACI) and Azure Kubernetes Service (AKS).

Storage Virtualization

Storage virtualization is a form of hardware virtualization (Preston 2002). Storage devices, potentially from multiple networked, physical locations, are grouped to appear as a single storage device or single storage location to an end-user or application. The virtual storage location is managed from a single console in most instances. Software applications implement storage virtualization.

Storage virtualization most frequently provides facilities for data and file backups, archival material, and disaster recovery. It hides the underlying complexity of the physical storage devices from users and ensures a simple interface. Figure 3.10 illustrates.

Storage virtualization generally is implemented using a combination of specialty hardware (appliances) and software applications. The goals are to: (i) enable and enhance storage management in environments where not all hardware is the same (e.g. heterogeneous); (ii) provide a managed system for equipment replacement and downtime reductions; and, (iii) offer better storage utilization. Several types of storage devices play important roles in storage virtualization. To understand how it all works, first it is important to differentiate between DAS, NAS, and SAN.

DAS (Direct Attached Storage)

Direct attached storage or DAS refers to storage devices directly attached to one computer. DAS is not meant to be networked and is used exclusively by one computer. Quite obviously, DAS does not make good cloud storage. Instead, DAS is intended for individual computer users and takes the form of a hard drive or a solid-state drive (SSD). From an enterprise computing standpoint, DAS can include a server's internal disk drives and external drives attached with SCSI or other interfaces. In this case, the drives may be accessible via a network but only through the server to which they are attached (see Figure 3.11).

DAS also comprises other devices such as optical drives, USB keys, DVDs, CDs, and so forth. DAS is dedicated to a single server and therefore storage is not pooled. It is limited in scale by the number of expansion slots in a server and other device constraints. DAS is inexpensive but less flexible and scalable than other forms of storage. Most physical servers boot from DAS. In part, this is due to the speed with which attached SSDs can operate. DAS devices are also vulnerable to failure, loss, and theft since they offer a single point of failure. On the other hand, they are not typically available on networks so that offers a measure of protection from particular security threats.

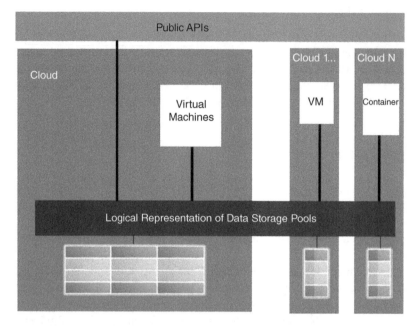

Figure 3.10 Virtualized storage masks the underlying physical devices from the users.

Figure 3.11 DAS example – computer has internal hard drive and an external drive attached with a SCSI connection for local use.

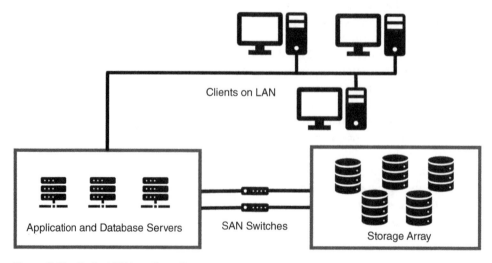

Figure 3.12 Typical SAN configuration.

SCSI: A Trustworthy Interface often used with DAS

In the days when storage was costly and slow, most application servers had internal storage devices. These storage devices used a protocol (e.g. convention for communicating) known as Small Computer System Interface or SCSI. SCSI is a standard to ensure devices can speak the "same language." Many hard drives, CDs, DVDs, and other storage devices can be attached using SCSI. SCSI interfaces require use of a microprocessor-controlled, smart bus, that permits peripheral hardware devices to be daisy-chained to a computer. In addition to storage devices, scanners, printers, and other items use SCSI. In the desktop and laptop markets, SCSI is losing out to USB and FireWire interfaces. In the server arena, SCSI remains popular, particularly for building in redundancy.

For instance, several hard drives can operate on a single server in a RAID (Redundant Array of Independent Disks) configuration. If a drive fails, a new one can be added without losing data while the system operates (at least from the user's perspective) as normal.

In addition to SCSI, several other interfaces are popular with DAS (and other forms of storage). Among these are SAS (Serially Attached SCSI), iSCSI (Internet Small Computer System Interface), and FC (Fibre Channel). Like SCSI, SAS uses the same basic command set to share data with attached devices. Since SAS uses a serial connection instead SCSI's parallel approach, it can increase throughput and transfer data at higher speeds. Also, SAS is a full duplex technology. This means it can send and receive data simultaneously. iSCSI moves SCSI onto the Internet. It uses an IP-based standard for connecting servers to storage devices. It is not truly a DAS technology but often is used for SANs and NASs. It provides a way to give a storage device location independence at a low cost. iSCSI relies on the server's CPU to process storage requests so this must be considered in overall system performance. Fibre Channel overcomes the speed constraints of iSCSI and enables high speed data transfer. Like iSCSI, Fibre Channel is used to connect storage devices using many different approaches.

SAN (Storage Area Networks)

SANs are used in business settings, particularly among SMEs. A typical SAN configuration comprises switches, disk controllers, host bus adapters (HBAs), and fibre or iSCSI connections. Figure 3.12 illustrate the general configuration of a SAN.

SANs enable users to share storage arrays in ways that are flexible, scalable, and ensure data redundancy. In general, SANs provide the ability to share storage arrays among application servers. SANs enhance performance and ensure data is safely and redundantly stored. SANs cost more and, if set up in-house, usually require specialized staff members due to their complexity.

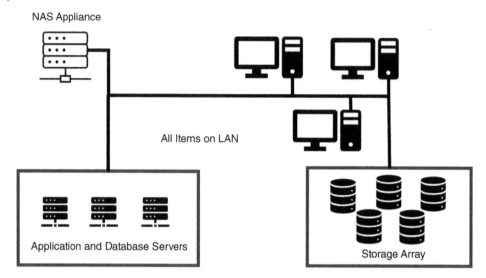

NAS Appliance

All Items on LAN

Application and Database Servers

Storage Array

Figure 3.13 Typical NAS configuration.

NAS (Network Attached Storage)

Network Attached Storage (NAS) is another form of storage configuration. NAS uses a dedicated piece of hardware, called an appliance, to add commonly accessible storage to a network. In larger, complex systems, a NAS appliance can help manage file requests, provide additional security, and offer methods for configuring and overseeing a shared storage system. Another unique feature of NAS is its use of file-level transfers as opposed to the block-level approach more commonly used in DAS and SAN approaches. In a way, it is like adding a large, commonly accessible hard drive to your computer, except that it is used by multiple computers. Unlike a SAN, NAS uses an existing network for the appliance and the storage devices. NAS often uses iSCSI for clients to interact with storage devices across a network and allows an IT specialist to configure these devices into a large, commonly accessible storage pool. Figure 3.13 illustrates.

A NAS approach for storage works well for SMEs with geographically dispersed offices running typical office or accounting software. In a sense, it can be used like a private cloud. In one scenario, each office could have its own NAS device and these devices could be connected centrally using a WAN. Software could be used to back up the remote offices from a central IT location, rather than at each individual site. This would facilitate disaster recovery efforts.

NAS Head

In some circumstances, NAS and SAN technologies can be synergistically combined. One way of doing this is through a NAS head device. A NAS head is a protocol handler gateway specifically engineered to interface with a SAN. A NAS head does not have any client storage availability, instead it provides a NAS-style interface that, from the end-user's perspective, makes storage appear to occur at the file-level. The NAS head offers the load-balancing and protocol handling needed to transparently change file-level access to more efficient, block-level format access. From the user's perspective, it looks like a giant NAS storage device. At the same time, IT specialists see the block-level data stored on the SAN so all their SAN management tools can be used to ensure back-up, recovery, storage efficiencies and other high-end features offered by a SAN. For an SME (or more likely larger businesses), it may be the best of both worlds. In many ways, this is a private cloud.

Storage Virtualization Techniques

DAS, SAN, and NAS develop pools of storage for use within an organization's IT infrastructure. Virtualization of storage pools makes sense to better use existing capacities and provide a method for rapidly scaling, should the need arise. Both NAS systems and SANs can be virtualized and several approaches exist for both.

SAN Virtualization: Virtualization can be applied at various levels within a SAN and to different storage functions (e.g. physical storage, RAID groupings, logical units and so forth). In general, storage virtualization can be broken into four layers:
1. Physical storage devices
2. Block aggregation layer
3. File/record layer
4. Application layer

A wide variety of hardware and software products exist to help consolidate multiple storage disks and systems into logical units grouped in ways that make sense to users, rather than relate to actual physical characteristics. This is particularly useful with systems composed of a variety of hardware and software types. SAN storage virtualization offers capabilities and advantages to an organization that includes access to all storage through a single central interface, and added ability to automate management, perform backups, expand storage, and update more easily.

NAS Virtualization: In a sense, NAS virtualization is creation of a private cloud for storage. Software products exist to enable an existing NAS device to abstract stored files from their physical location to an online virtual location, almost like a private Dropbox application. Rather than rely on a LAN or WAN for accessing files, a virtualized NAS device shares or streams files over the Internet to any computer with appropriate privileges to permit use. This is particularly helpful to organizations or individuals that consume stored media or need to have mobile devices connected to organizational storage depositories. A virtualized NAS device offers the same benefits as a private cloud: no third party is involved, and data remains private on the organization's own storage devices behind a firewall. Virtualized NAS devices can be attached through virtual private networks (VPNs) for added security.

File- Versus Block-Level Virtualization

Storage virtualization can be considered in terms of file-level versus block-level. Remember that SANs (and DAS) use block-level virtualization and NAS systems use file-level storage. Each of these approaches is virtualized in somewhat different ways. Block-level storage virtualization adds a layer to the SAN which can be thought to sit on top of the storage arrays. The servers using the SAN are directed to virtualized logical unit numbers (LUNs). In SANs, LUNs are unique identifiers assigned to each device. Virtualization at this level causes separate physical devices to be logically "regrouped" using LUNs that make the hosts see the locations differently. The physical devices are transparent and now the logical units can be accessed using the appropriate protocol (perhaps iSCSI or Fibre Channel). So, the virtualization is essentially a translation layer between hosts and storage arrays in the SAN. Servers access virtualized LUNs instead of the LUN's originally assigned within storage array. File-level storage virtualization, usually done with NAS devices, virtualizes the link between the files and their physical storage location. This means that logical drives, composed of various physical storage areas are seen by the users. The actual physical locations remained masked. File-level virtualization enables location independence, and this can enhance a user's capability to move and access files in user-friendly ways that make more sense.

Summary

Virtualization is a primary characteristic of cloud computing. Hardware virtualization, operating system virtualization, operating-system-level virtualization (containerization), and storage virtualization are all examples of approaches to add a logical layer that redefines, regroups, and consolidates resources in ways that add efficiency, management capabilities, and expands usage possibilities. Virtualization ensures that resources are shared more effectively, and that expansion can occur rapidly and without user impact. More importantly, it often means preempting end-user confusion.

References

Barrett, D., Kipper, G., and Liles, S. (2010). How virtualization happens. In: *Virtualization and Forensics: A Digital Forensic Investigator's Guide to Virtual Environments*, 3–24.

Portnoy, M. (2016). *Virtualization Essentials*, 2e. Hoboken, NJ: Sybex.

Preston, C.W. (2002). *Using SANs and NAS: Help for Storage Administrators*. Sebastopol, CA: O'Reilly Media.

Further Reading

Jain, N. and Choudhary, S. (2016, March). Overview of virtualization in cloud computing. In: *2016 Symposium on Colossal Data Analysis and Networking (CDAN)*, 1–4. IEEE.

Kapil, D., Tyagi, P., Kumar, S., and Tamta, V.P. (2017, August). Cloud computing: Overview and research issues. In: *2017 International Conference on Green Informatics (ICGI)*, 71–76. IEEE.

Manco, F., Lupu, C., Schmidt, F. et al. (2017, October). My VM is Lighter (and Safer) than your Container. In: *Proceedings of the 26th Symposium on Operating Systems Principles* October 28, 2017, Shanghai, China, 218–233. ACM.

Sacks, D. (2001). Demystifying storage networking DAS, SAN, NAS, NAS gateways, Fibre Channel, and iSCSI. *IBM Storage Networking*: 3–11.

da Silva, V.G., Kirikova, M., and Alksnis, G. (2018). Containers for virtualization: an overview. *Applied Computer Systems* 23 (1): 21–27.

Tate, J., Beck, P., Ibarra, H.H. et al. (2018). *Introduction to Storage Area Networks*. IBM Redbooks.

Wüst, S., Schwerdel, D., and Müller, P. (2017). Container-based virtualization technologies. *PIK-Praxis der Informationsverarbeitung und Kommunikation* 39 (3–4): 51–61.

Yadav, A.K. and Garg, M.L. (2019). Docker containers versus virtual machine-based virtualization. In: *Emerging Technologies in Data Mining and Information Security*, 141–150. Singapore: Springer.

4

Can the Cloud Help Operations?

Organizations rely on IT infrastructure to ensure their daily operations achieve organizational goals. This can be tricky since infrastructure often must be put into place anticipating future needs and demands. In traditional IT systems, this meant taking an educated guess and making very expensive capital investment purchases–mainframe computers, storage capacity, software purchases and leases, networks, and so forth. The cloud has eased many of these pressures by introducing an appropriate infrastructure that might be informally called, "pay-as-you-go." Several cloud features provide flexibility to organizations and help keep operations on track without the upfront investment. Two of the tools that help with this are load balancing and scalability/elasticity (Ahmad and Andras 2019). This chapter provides a look at both in more detail.

Load Balancing

Load balancing is the process of efficiently distributing incoming network traffic among servers, also called nodes. In larger operations, multiple network servers, known as a server farm, server pool, or backend servers, are used to prevent the system from being overwhelmed with a high volume of requests.

Load balancing algorithms are classified according to various criteria. In general, they can be static or dynamic, and centralized or distributed. Static load balancing is a pre-determined approach that depends on a set of rules, or the system load when the node is selected. Dynamic balancing, on the other hand uses current load information for request distribution while the system is running. Dynamic algorithms are adaptive and change as the system state changes. In general, systems expecting constant, uniform requests work well with centralized, static load balancers. On the other hand, unpredictable, varying loads often are better served with nodes using a decentralized, dynamic algorithm. Theoretically, more information collected at run time results in better node selection decisions. But the trade-off becomes more system communication which can slow processing overall.

Centralized algorithms rely on a control management node used to gather information from all nodes eligible to process loads. The central node makes the distribution decisions. In distributed algorithms, no central manager exists. Instead, nodes eligible for processing

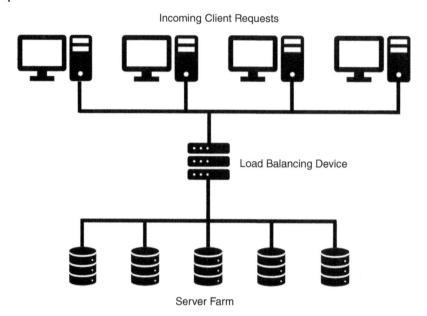

Incoming Client Requests

Load Balancing Device

Server Farm

Figure 4.1 Client requests are sent to various servers to avoid one from being overwhelmed.

exchange their system state information with every other node. As mentioned earlier this inter-node communication can slow processing.

In centralized approaches, a hardware or software load balancer sits between the clients and server farm performing a "traffic cop" or "director" role. Requests are allocated across all available servers in an intelligent way. This ensures all servers are utilized to maximize speed and enhance utilization. Servers are used uniformly to extend their lifespan. If one server goes down, the "traffic cop" redirects traffic to other servers. Likewise, if a new server joins the pool, the "traffic cop" reassesses its algorithm and seamlessly changes the traffic pattern.

Put into context of modern, cloud-based systems and websites, imagine high-traffic resulting from interaction of hundreds of thousands (or more!) clients, all making concurrent requests to web servers. Imagine the server pool needing to interact with a SAN to retrieve text, images, and videos that must be sent quickly and reliably to the requesting clients in an organized way. Figure 4.1 shows a conceptual view of how a centralized load balance device determines which client request is sent to each server to prevent one machine from becoming overwhelmed.

Load balancing is a key feature of cloud computing. The concept of load balancing is extended to distributing server workloads and computing resources all within a cloud environment. Further, load balancing can be more complex and sophisticated than a "traffic cop" pointing to various servers. IT specialists may build special business rules that create priority requests, allocate servers to clients, and manage application and workload demands in a variety of ways. In general, a load balancer accomplishes the following:

- Ensures client and application resource requests (as well as network traffic load) are efficiently allocated across multiple servers.

- Manages resources to ensure high levels of uptime by monitoring availability and sending requests to online servers within the pool.
- Ensures flexibility of resource pool size and provides tools for elasticity and scalability according to demand levels.

In cloud environments, particularly those that are highly virtualized, nodes may be created or removed as required to help balance loads. This can become very complex and involve sophisticated data collection and processing (Chaudhury et al. 2019).

Load Balancing Algorithms

Many strategies and approaches are used to implement load balancing. Several general approaches exist as do many vendor-specific approaches meant to ensure the superiority of their products. To start with, load balancing algorithms can be viewed as either static or dynamic. We will look at both approaches with a focus on key algorithms. Keep in mind that many others exist besides those covered.

Static Load Balancing Algorithms

Static load balancing algorithms are determined upfront by IT specialists and then put into place for use. In a sense, these are best practice approaches that may not use specific network characteristics to determine or adjust load allocation. These methods are classified as non-pre-emptive. That means once a request has been allocated to a specific server, it will not be transferred to another. A primary motivation for using a static approach is reduction of execution time. Less server monitoring and communication takes place. Therefore, algorithm execution time is greatly reduced. A primary downside to this approach is that the current system state is not considered when allocations are made. That means if a server is slowed, it may still receive additional workload. Overall system performance may suffer due to this shortcoming, and requestors may experience noticeable differences in response times from visit to visit.

To illustrate, visualize the example of a customer service call center. A static algorithm could mean that each customer calling for help is allocated to a specific service representative in order. The algorithm does not consider whether a representative is on a one-minute call or a 60-minute call; or if she is busy or idle. Each of the representatives gets the same number of customer service calls and has their own individual queue. In this case, some customers may have a longer wait than others. In a dynamic algorithm, the representatives' current states (e.g. currently busy or not) would be considered. If a representative was busy, then the next call would be sent to the next idle representative rather than being put into a queue. While this seems to make more sense, the system does have to determine if each representative is busy or idle before routing the customer's call. In a situation where all calls are very short with similar requests, the added step of determining whether a representative was idle or not would be extra activity and a static algorithm would work fine instead, especially when very large numbers of representatives are involved. In a situation where customer calls vary in length and complexity, it would make more sense to have an algorithm that is more sophisticated in choosing the representative to receive the next caller.

If a static approach is used, a variety of static load balancing algorithms are available. Among these are random, round-robin, and IP-Hash.

Random: Randomly selecting the next server to receive a request is one method used to distribute workload. This approach does not require information about the last server selected nor does it attempt to determine current loads on each server. Random selection means that over time every server will receive approximately the same number of requests. It does have shortcomings. For instance, sometimes a server may become overloaded while others may not have any load at all. It is possible that the same server may be randomly selected multiple times in a row. Therefore, this approach should be viewed as providing long term equality at the expense of potential short-term issues. This algorithm works best when the servers are homogeneous, and the requests are of approximately the same magnitude.

Round Robin Algorithm: Again, this static approach to node selection is best suited when servers are homogeneous, and requests are approximately the same magnitude. In this scenario, server requests are put into an ordered sequence. Requests are sent to the servers in order regardless of server load or whether a server is overloaded or not. When the last server is reached, the next request goes to the first server in the sequence and starts over. This algorithm results in a load that is distributed to all servers equally. Round robin is an easy algorithm to implement and requires no communication between servers nor does it require knowledge about its servers' states. But, if requests vary in length and complexity, a given node could become backlogged.

IP Hash: This algorithm is very similar to the random method. However, instead of selecting the next server randomly, the IP address of the client is used to select the host to service the request. This distributes load somewhat randomly. Like the random approach, it does not require any information about the last server selected nor does it attempt to determine server load levels. As with the random approach, a server may become overloaded while others may not have any load at all. This algorithm works best when the servers are homogeneous, and the requests are approximately the same magnitude.

Dynamic Load Balancing Algorithms
Dynamic load balancing algorithms use available information when trying to decide what server should receive an assignment. These algorithms monitor each servers' status and related changes within the pool to distribute work and make assignments. Centralized dynamic algorithms rely on a central node for communication and load assignment. Decentralized dynamic algorithms move the decisions down to the nodes themselves. Three general algorithms used for dynamic balancing are central, transfer, and least connection. While many more exist, we will focus on these.

Central: This approach, as the name implies, uses a centralized and dynamic approach. When the central manager receives a request, it is stored in a FIFO (first-in-first-out) queue. The requests remain in the queue until one of the processing nodes sends a notification that it needs more work. The oldest request is removed from the queue and work is sent to the node. If no work is in the queue, the request for more work is stored until a request comes. Then the central manager makes the assignment. This approach requires communication between the processing nodes and the central manager. Think back to the customer server center example mentioned earlier. Imagine that instead of requests being sent round-robin to each customer service representative, each with their own queue for pending requests, that now a single queue serves all representatives. When one task is finished, a new job is pulled from the central queue and sent to the representative that has been idle the longest. This is a better way to assign work, but it does require that the central manager knows when the representative finishes each task and hence increases internal communication and processing requirements.

Transfer: This algorithm looks at on-going server loads and determines if one node is over-worked. If so, the central manager can decide to move a task from one server to another for processing. Basically, this algorithm was created to manage inter-process task migration to help keep all nodes operating at relatively similar levels. The algorithm can be implemented in different ways. From a centralized perspective, the central manager monitors the node workloads and makes the decisions. When using a decentralized approach, processing nodes that become overworked can request help from other nodes that may not currently be as busy. The distributed approach requires greater levels of internode communication.

Least Connection Algorithm: This algorithm looks at load distribution based on the number of active connections between a server and its clients. Distribution of new loads (or re-distribution of tasks) is influenced by rules for keeping the number of connections approximately even. To use this algorithm in a centralized manager environment, the load balancer tracks the numbers of connections per node. A node's number increases for each new connection and decreases when a connection is finished or times out. The central manager sends tasks to the least connected nodes. In a decentralized approach, nodes with too many connections essentially send out a request for help to other nodes that may be less busy.

Cloud Load Balancing Algorithms

In cloud environments that are highly virtualized, additional load balancing algorithms may be used. In these situations, virtual server nodes can be created as demand increases. Interesting ideas and implementation algorithms include variants of the central manager and threshold approaches. Other more sophisticated and imaginative approaches also exist. Among these are hybrid, ant colony, bee's life, and genetic algorithm.

Central Manager: This balancing algorithm is used in the cloud as well as in traditional client–server networked systems. In the cloud, just as on a traditional system, a central server selects a node for load transfer. Generally, the server having the least current load would be chosen to receive the newest request. The central manager is responsible for maintaining an index of all potential processing nodes. As a node's status changes, it notifies the central manager. The goal is to send the least busy server the next job and keep work flowing evenly. As with the traditional central manager algorithm, this one is subject to overwhelming the central manager with messages, particularly on very large, cloud-based systems. In some circumstances, new processing nodes are created as load increases, or nodes are removed if load levels reduce.

Threshold Algorithm: This algorithm takes advantage of cloud computing power. As demand increases, new processing nodes are created. Loads are assigned immediately to these new nodes. Node selection is done locally, in distributed fashion without sending messages to a central manager. Each node has its own copy of the system's load and this determines request assignments among existing nodes. Load is viewed in one of three states: underloaded, average loaded, and overloaded. New nodes are initiated in the underloaded state. After task assignments, and when the node's work level exceeds a pre-determined level limit, it sends messages about its state change to all active, processing nodes. When enough tasks are completed, the state changes back to underloaded and new tasks may be acquired. In general, this algorithm has lower overhead and better performance than the central manager approach.

Hybrid: Many hybrid load balancing algorithms exist. In general, these algorithms use multiple stage approaches to determine the best node to receive a processing task. For example, a two-stage model may consist of these stages: (i) perform a static balancing algorithm; and (ii) enhance node selection with dynamic analysis. For example, if step (i) uses a round-robin algorithm, it may select the next node to receive a task then in step (ii), check its current workload and processing status. If it is operating above a pre-determined threshold, the node may be skipped in this iteration and the two-step process repeated for the next node. Rather than rely on messaging, the central manager in this approach may use an agent to gather node operating information which could include values such as memory capacity and CPU usage level.

Ant colony: There is a constant search for better ways to balance loads in cloud-based systems. Among several very imaginative methods are ones based on the natural world. The ant colony approach views the server pool from the perspective of worker ants. The ants are intelligent agents that move forward and backward to track overloaded and underloaded nodes. Ants update "pheromones" to track node resource information. Two update types are: (i) foraging pheromones which help find overloaded nodes; and, (ii) trailing pheromones to help find underloaded nodes.

Bee's life: Like the ant colony algorithm, this one is inspired by the natural world and a honeybee's process of searching for food. The bee manages loads across VMs in ways that reduce task queue waiting times (Dinesh Babu and Krishna 2013).

Genetic algorithm: A third natural world motivated algorithm uses a genetic algorithm approach to optimize the node selection process. This algorithm employs a "survival-of-the-fittest" paradigm to ensure algorithms with higher efficiency produce "offspring" that may inherit their best features over a series of generations.

Different load balancing algorithms are developed with various motivations and benefits in mind. Simply put, different approaches work better in different environments which vary regarding homogeneity of server processing nodes, the nature of processing tasks and requests, and the priorities/needs of business outcomes.

Client-Side Load Balancing in the Cloud

A few of the balancing algorithms covered in the prior section alluded to clients being able to ask that work be off-loaded. Another approach, called client-side load balancing, does not refer to the end-user as the client, but rather middleware acting as a client for back-end services. Just as a refresher, remember that middleware is software that acts as a bridge between backend services like a database, and users' applications. Front-end services interface with the user. Client-side Load Balancers (CLB) use an elastic cloud storage service to choose back-end databases or web servers based on pre-determined criteria like current load, queue length, average processing time or other factors. In this case, the client finds the best service instead of a central manager or agent doing it.

Hardware Versus Software Load Balancing

In traditional IT operations, load balancing relied on hardware solutions. Generally, this hardware resided in an on-premise data center maintained by in-house IT specialists. Balancing used a special appliance from a vendor like Cisco, Citrix, or Barracuda. These specialized devices sat in server racks and distributed traffic to directly connected physical servers.

At first, load balancing generally was done by large companies with a substantial IT staff and budget. The cloud enables a new family of solutions for SMEs. Rather than rely on hardware-based balancing, most cloud providers use software solutions running on virtual machines. Software balancers often are part of the server suites included by cloud providers. This makes the technology affordable even for small companies.

NGINX Plus

This load-balancing software, used by many leading cloud-based organizations, helps with processing requests. High-volume sites like Netflix and Dropbox have used this cloud load balancing solution to ensure content delivery in a secure, reliable fashion.

Most cloud vendors offer load balancing software or software/hardware configurations. It is no surprise that Amazon, Microsoft, Google, and Rackspace have products and services in this area. For example, Amazon has AWS Elastic Load Balancing to distribute processing requests among various instances. Google has a balancing service included as part of its IaaS servers. Azure includes Traffic Manager to help distribute load among data centers

Cloud-Based Balancing

Many IT specialists consider it best practice to provision the load balance server in the same environment as what it balances. This means, for an organization using the cloud, cloud-based balancing is preferable. In addition to balancing loads, cloud-based systems must balance goals. One is to ensure traffic and load are routed in efficient and sensible ways. Another is to maintain security. This recalls a common problem known as *noisy neighbors*.

Cloud environments provide cost savings, elasticity, and other benefits in part due to their multi-tenant architecture. Like an apartment complex, if the walls are thinly insulated or if too much infrastructure is shared (e.g. hot water, Internet connection, air conditioning and so forth), the presence of the annoying neighbor becomes more obvious. In cloud environments, the noisy neighbors may monopolize bandwidth, storage access, CPU, and other resources. This means your system performance may be impacted by what your neighbors are doing! Imagine if you were taking a shower and when your neighbor flushed her toilet, your water suddenly became too hot. Ouch! In the cloud, a weak architecture can impact virtual machines and applications.

What causes noisy neighbors? Again, like an apartment complex that has been put together with cheap walls, windows, doors, plumbing, and other amenities. A "cheap" cloud means not enough infrastructure exists to accommodate all the tenants. In a private cloud environment, this might mean the underlying hardware infrastructure is insufficient and needs upgrades. In a public cloud it could mean a lack of bandwidth causing degraded network performance. Remember, bandwidth must be sufficient to carry all network data. So, low bandwidth will slow speeds and latency (response to requests) especially when serving Websites, databases, storage and microservices.

In a private cloud, noisy neighbors are avoided by ensuring adequate bandwidth is present and by having up-to-date fast storage devices and servers. If rapid response is paramount, then more physical servers can be added to the server pool. Having an effective balancing algorithm will ensure requests receive prompt fulfillment by moving workloads to the best physical servers.

If the situation is mission critical, a private cloud can use a bare-metal approach. This means each application runs directly on its own assigned hardware. In a sense, this is a single-tenant environment. Think of it as an Airbnb rental where you get an entire house rather than a room in a house that is shared. Of course, a bare-metal solution is more expensive and negates some broader benefits offered by cloud computing. But organizational needs often are unique and sometimes a single tenant solution will be required.

In public clouds, a bare-metal solution may not be possible. To help avoid these situations, cloud vendors have developed *virtual local area networks (VLANs)*. VLANs partition a network into components which match the business needs of those using computing resources. For instance, the VLAN might be partitioned by functional business department, or by required security levels. VLANs can divide a system into logical groups and establish related rules about use. VLANs can be simplistic and mimic a traditional office environment with printers, shared drives and so forth. Or, they can be highly complicated to meet security clearance and legal requirements of a business or industry. A VLAN can also mimic geographic configurations if that has an advantage.

Cloud Load Balancing Versus DNS Load Balancing

In a discussion about load balancing, you may hear the phrase, *DNS balancing*. DNS balancing is a network optimization technique. This form of balancing is used to route a web domain's incoming traffic flows to an appropriate web server. Cloud load balancing has its roots in DNS balancing but can be very different and more comprehensive. DNS load balancing is concerned with faster access to resources served back to a client connected to a website. It provides multiple IP addresses for the same host and routes traffic among different servers. DNS load balancing distributes load requests for a domain.

Cloud load balancing takes the idea much further. Instead of transferring requests to a pool of web servers providing services to the same domain, it routes among data centers, server pools, virtual machines, and/or other resource groups. Cloud load balancing is much broader and meant to handle a wider variety of traffic and workload needs.

Scalability and Elasticity

Scalability and *elasticity* are terms often used interchangeably. In general, both relate to ensuring computing resources match well with resource demand (Erl et al. 2015). As such, scalability and elasticity are the primary features and benefits of cloud computing.

Figure 4.2 Scaling "out" using guards as an example.

Although similar, they are different. Elasticity describes methods used to adjust a system on-the-fly to match workload changes. This can be up in scale or down in scale. Scalability is specifically related to the system's capability to accommodate larger loads easily by adding resources. This might mean improving hardware power (scale up) or adding additional nodes to the server farm (scale out). Scalability is the capability to grow, elasticity is the ability to react to changes dynamically. Scaling may not always happen automatically.

For instance, scaling up is increasing individual capabilities. A server may be scaled up by adding more memory or more storage. Scaling out refers to duplication and expansion by adding more. This would mean adding more servers to the pool. Think of an example in terms of a military group. Scaling up would mean an individual soldier receives training so they can perform better. Scaling out would mean instead of having one soldier on guard duty, four were put on duty. Elasticity would mean adjusting the number of soldiers assigned to guard duty based on the currently perceived threat level. See Figure 4.2.

Since elasticity is dynamic, it is a more important feature of cloud computing environments where resources are tied to cost. Both elasticity and scalability are managed by business rules specified by an IT system manager or corporate policy. In the case of elasticity, these rules provide a basis to provision and de-provision resources automatically. The goal is to match resource availability with resource demand at every point in time. Since most cloud-based systems operate on a pay-as-you-go manner, elasticity often has economic efficiency as an essential motivator. It also ensures, in cloud environments, that resources not being used by one subscriber, can be used by another. This means the cloud provider leverages their hardware investment in the best possible manner to reduce everyone's costs. So, elasticity is defined as dynamic adaptation of computing resource capacity. These resources may include computing power, storage space, application subscriptions, database capacity, and any other cloud-based resources.

Scalability on the other hand is not quite so dynamic. It might mean that system resources can be increased when the right time comes but it may not be an automatic action. Instead, it may involve purchase of additional resources or other changes. While scalability is important in computing environments, elasticity is a better feature in cloud-based systems. Therefore, we find the term scalability often used in managerial level business settings while elasticity is used more by technical people and IT managers.

Elasticity in Cloud Environments

A primary way elasticity is implemented in cloud environments goes back to the concept of using a hypervisor to manage VMs. If you recall, a hypervisor is essentially a virtual machine monitoring layer used to create, run, and oversee virtual machines. In the context of the current chapter, think of the hypervisor being equivalent to the central manager

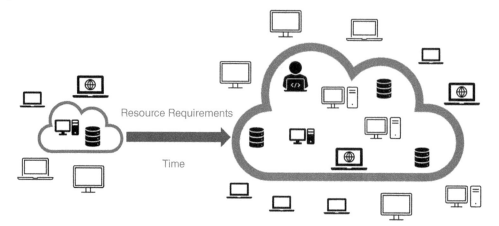

Figure 4.3 Elasticity means up and down resource provisioning as needed. Source: Modified from Tango!-Project/Layout: Sam Johnston 2011.

in load balancing algorithms. In cloud infrastructure, the hypervisor has the capability to create new VMs or containers as system demand ramps up. Likewise, as demand decreases, the hypervisor can remove extra VMs or containers. So, the hypervisor works in real time to ensure computing resources meet demand. This sounds perfect, but challenges and problems do exist.

Challenges for Elasticity

Elasticity is an amazing feature for cloud computing that provides huge benefits for cloud service providers and client service users. However, a broad-reaching technology such as this poses technical and managerial challenges. We look at several of these including learning curve and implementation, response time, application monitoring, stakeholder needs, multiple levels of control, security, and privacy/compliance. See Figure 4.3.

Learning Curve

Implementing a cloud-based IT infrastructure, particularly in environments that once relied on traditional IT infrastructure, comes with a learning curve. Changing to an elastic cloud requires rethinking the ways resources are allocated, costs are accounted, and systems are licensed and used. For example, in cloud-based environments, traditional roles such as administrator and developer overlap. Rethinking what is automated and what should be monitored is another part of the challenge, especially since each cloud environment is unique. So, key IT personnel need training and must understand how to serve their organizational end-users.

Most cloud vendors offer online courses to help IT personnel get up to speed. In addition, many universities and training consultants provide services. Best practices, processes, technologies, and many other topics are covered in these offerings.

Response Time

A big challenge with elasticity is time. Adding a VM or container ensures user demand is met, but in a high-speed environment, implementing elasticity requires time. For instance, a cloud VM could be required at any time by an end-user. However, creating a VM is not instantaneous. In fact, in current environments, it may take a hypervisor several minutes to provision the VM, so it is ready for a processing request. Elasticity does not just involve managing server requests. It could be the request for a new virtual desktop needed by an onboarding employee. Or, it could be a new "seat" for a software package licensed to an organization. The time to implement a new VM is dependent on many factors, such as image size, the type of VM required, the number of current VM requests, and data center characteristics. So, elasticity takes time but as technology and computing speeds continue to improve this will become less of an issue.

Monitoring Elastic Applications

A technical challenge faced when using elastic cloud concepts includes approaches, rules, and ways that applications and users trigger resource allocation and deallocation. As mentioned earlier, these resources might include more storage space, additional VMs, new desktop instances, additional computing power and so forth. Cloud elasticity is inherently volatile and that means tools monitoring resource use need to be sophisticated. For instance, a monitor might suddenly indicate the system is using too many resources when in fact this was desired. Or, perhaps a cost cap is in place to prevent IT expenses from exceeding a threshold and if new VMs are added at a crucial time, this cap may need to be changed by an administrator. Other more technical problems include how VMs are monitored when they might suddenly disappear, particularly if a tool is looking at their cost; and, what aggregation metrics should be used for accurate cost calculations. Fortunately, these problems are not unique to one client organization so cloud providers have solutions. It is just important to know that new capabilities often bring new challenges.

Stakeholder Needs

Like the monitoring issues, many other stakeholder challenges exist. For instance, requirements of each need to be gathered and considered during the implementation process to ensure elasticity rules are properly put into place. Stakeholder performance needs must be weighed against costs. Considerations include how fast a new VM needs to be put into play versus how much money that speed costs. Expectations must be realistically considered prior to implementation and use.

Multiple Levels of Cloud Control

Cloud computing, by its nature, serves the needs of many users. This means many structures must be in place to control access to resources and ensure the right people can connect to the right systems. Likewise, data access must be controlled. Control strategies can become complicated when containers, VMs, and applications are woven together in a complex way. If the strategy is not well considered, conflicts between various levels of controls may emerge. While most cloud providers supply a variety of control levels and privileges, setting this up and maintaining it, particularly in an elastic environment where new resources may automatically appear, must be well considered.

Security

Related to control levels, general security becomes a major concern for cloud implementations. Elastic environments mean resources appear and disappear and that access to these resources must be managed carefully. Organizations do not own the hardware so access to the server room is less of an issue. Cloud providers are responsible for physical security. Likewise, virus protection, data security and other traditional concerns are managed by providers. The organizations using the cloud services still must manage application use and user authentication practices. So, this remains a challenge in cloud environments.

Privacy and Compliance

Data privacy and compliance with international laws (which most modern organizations must now consider) is a challenge due to its inherent complexity. Cloud providers offer many services in this area and manage a portion of the responsibilities. But both cloud service providers and client companies must be knowledgeable about the laws that apply to their operations. For instance, European data privacy laws have a very different focus than the laws in the United States. China has its own conventions and laws as well. In addition to following laws and regulations, cloud vendors provide facilities for audit logs and trails. Audit teams are in place at most major cloud providers and these teams can help a cloud services client set up any logging functionalities required by law.

Benefits of Cloud Elasticity

Of course, elasticity would not be a primary feature of cloud computing if it did not offer substantial benefits. The idea that VMs or containers can be automatically created or removed as demand requires, and this automation can ensure an organization only pays for its computing consumption, is revolutionary. It eliminates the need to guess at future capacity. Likewise, many issues associated with equipment and software obsolescence are removed and accounting practices can tie use to the organization's departments that consume the resources. Among the benefits of elasticity in cloud environments are implementation ease, failover and fault tolerance, on-demand computing, pay-as-you-go computing, and standardization of server pools.

Ease of Implementation

Elasticity features are offered as services that can be purchased from cloud vendors. In some cases, elasticity is transparent and will just occur as needed. For instance, data storage needs, CPU requirements, and network traffic management may not need any oversight from a cloud client's IT staff. These items will ramp up as needed according to the agreement with the cloud provider, and the client company will be billed accordingly. Of course, a private cloud implementation will require on-premise IT staff to set automation and policies according to business needs. For items that do require a closer eye, in some cases container creation or VM deployment, the cloud vendor will provide expertise and the system can be configured to meet organizational needs. Best practices have been developed by cloud vendors to help with these situations. Client organizations can subscribe to training for its cloud administrators and IT staff members if more complex operations are required.

Failover and Fault Tolerance

The cloud provides a great deal of flexibility when it comes to failover and fault tolerance. A server that is failing can be cloned proactively and deployed on a new VM before the old one stops. This use of cloud elasticity properties can be automated and put into play without human intervention. This practice is particularly effective and fast if VMs are identical. Many cloud providers have a goal of making all organizational servers identical. This practice attempts to get IT managers to change their viewpoint on servers from that of *pets* to *cattle*. Pet-type servers are highly customized with hard-coded processes and specialized features. Pet-type servers are almost impossible to replace if something goes wrong. Cattle-type servers, on the other hand, are uniform using standardized features. That makes them easy to duplicate and replace with no extra setup time. A best practice in cloud computing takes advantage of elasticity by rethinking server deployment and replacing pets with cattle. Many benefits result from that transformation.

On-Demand Computing

Traditional IT infrastructure had no good way to deal with spikes in demand for computing, storage, or network resources. About the only way to manage this was to ensure excess capacity was available. So more hard drives, more CPUs, and more network capacity than usually needed had to be purchased and maintained. Of course, having excess capacity for the rare occasions it might be needed could be very costly. But not having it could be far worse with dire, long-term consequences. As a result, extra hardware had to be purchased and deployed. Also, this meant having extra space in the server room or data center together with enough power, cooling and back-up capability. And having the staff to run the extra hardware. The cloud offers an easy alternative with elasticity so when usage requirements spike, the capacity is readily available with cost only related to use. Cloud environments have eliminated most worries related to unpredictable use patterns. On-demand cloud computing's elasticity feature sometimes is called *autoscaling*.

Autoscaling

Cloud computing best practices include several methods for managing the elasticity feature called *autoscaling*. Most often, existing servers are watched or monitored. If an important usage threshold is reached, the system expands and adds a server. Autoscaling can also be based on business rules. For instance, in advance of "Cyber Monday" (e.g. in the U.S., this is the Monday after Thanksgiving when massive sales become available via Internet retailers) an organization might ramp up its server capacity in anticipation of greater demand on its website. A related approach is to use scheduled scaling to deploy server resources based on regular high demand periods. For instance, a customer service organization may do most of its work from 8 a.m. to 5 p.m.

Pay Only for What You Use

Tied closely to autoscaling is perhaps the underlying benefit that drives many organizations to a decision to use cloud computing: *pay-as-you-go*. The economic benefits of paying only for the computing, storage, and networking resources an organization uses is a huge benefit. Budgets become easier to develop. Economic cost-benefit analyses are more obvious and so forth. An organization no longer has idle IT resources waiting in case more computing or storage needs arise in the future. An organization does not need to

Figure 4.4 Archetype computer geeks working in the basement and hoarding knowledge may be less realistic in today's cloud computing environments. Source: FayFotoImage taken from https://commons.wikimedia.org/w/index.php?curid=32005874. Licensed under CC BY 1.0.

buy servers for short-term testing or development projects. Resources can be provisioned, used, paid for, and de-provisioned, and the budget for the projects is predictable and useful to business planners.

Standardization of Server Pool

In traditional IT systems, a common practice was to have a specialist on hand that understood the IT ecosystem including the hardware setup, software deployment quirks, and configuration settings. Ideally this knowledge would be carefully documented, but often, the specialist was not motivated to document all the nuances, instead opting to be the smart IT guy or gal that could just fix problems that came along. Figure 4.4 shows an archetype IT person in a basement full of hardware.

Hence emerged the stereotypes of the basement-dwelling IT geek, brilliant but lacking in communication skills, working late at night in the dark to keep systems humming along. This specialist was insulated from the rage of upper management because they could not do without them. Today, the concept of *infrastructure as code (IaC)* has emerged. This is a self-documenting process that ensures anyone with programming or scripting knowledge can understand system setup and infrastructure. IaC uses scripting languages such as XML, JSON, or Python to document system configuration. Common coding language usage is another benefit of cloud elasticity. We will examine this concept in more depth in Chapter 5.

Summary

Two concepts are at the root of cloud computing: load balancing and elasticity/scalability. Cloud load balancing ensures workloads requiring computing-related resources are distributed in an efficient and effective manner. Costs are reduced and resources shared using specialized algorithms to distribute the load. Further, cloud load balancing maximizes computing resource availability.

A variety of resources are often load balanced in the cloud including processing capability, network interfaces and services, application instances, storage acquisition, and more. Generally, cloud load balancing is done using software solutions from cloud vendors. Large-scale public clouds provide balancing features to ensure that applications maintain high availability and performance across multiple virtualized application servers.

Cloud load balancing relies on the concept of elasticity to manage demands as resources grow and shrink. Elasticity reallocates resources to scale up to meet increasing demands or scale down as resource needs decrease. Having an elastic pool of resources with dynamically changing allocations to various applications and users offers cloud vendors flexibility to serve the changing needs of their clients. Further, clients only pay for what they need to use. Large pools of idle, just-in-case capacity is no longer required. Another way to look at this is that elasticity is the capability to dynamically change computing resources to match demand. Scalability is the ability to upsize operations from a managerial perspective. Elasticity can make it happen.

References

Ahmad, A.A.-S. and Andras, P. (2019). Scalability analysis comparisons of cloud-based software services. *Journal of Cloud Computing* 8 (1): 1–17.

Chaudhury, K.S., Pattnaik, S., Moharana, H.S., and Pradhan, S. (2019). Static load balancing algorithms in cloud computing: challenges and solutions. In: *International Conference on Soft Computing and Signal Processing*, 259–265. Singapore: Springer.

Erl, T., Cope, R., and Naserpour, A. (2015). *Cloud Computing Design Patterns*. Upper Saddle River, N.J.: Prentice Hall Press.

Dinesh Babu, L.D. and Krishna, P.V. (2013). Honey bee behavior inspired load balancing of tasks in cloud computing environments. *Applied Soft Computing* 13 (5): 2292–2303.

Further Reading

Al Nuaimi, K., Mohamed, N., Al Nuaimi, M., and Al-Jaroodi, J. (2012). A survey of load balancing in cloud computing: challenges and algorithms. In: *2012 Second Symposium on Network Cloud Computing and Applications*, 137–142. IEEE.

Becker, S., Brataas, G., and Lehrig, S. (eds.) (2017). *Engineering Scalable, Elastic, and Cost-Efficient Cloud Computing Applications: The CloudScale Method*. Springer.

Ghomi, E.J., Rahmani, A.M., and Qader, N.N. (2017). Load-balancing algorithms in cloud computing: a survey. *Journal of Network and Computer Applications* 88: 50–71.

Mishra, S.K., Sahoo, B., and Parida, P.P. (2020). Load balancing in cloud computing: A big picture. *Journal of King Saud University-Computer and Information Sciences* 32 (2): 149–158.

Sharma, H. and Sekhon, G.S. (2017). Load balancing in cloud using enhanced genetic algorithm. *International Journal of Innovation and Advancement in Computer Science* 6 (1): 23–29.

5

How Are Clouds Managed?

Cloud computing is possible due to amazing innovations in technology. But equally important are facilitating tools and techniques developed by cloud vendors to enable implementation and day-to-day operations. Among these management features are automation, orchestration, replication, and disaster recovery as a service (DRaaS). This chapter explores concepts related to these topics and provides examples of how the "ops" side of DevOps has made innovative strides in recent years.

Automation

Cloud automation is a catch-all phrase used to describe the processes and tools that replace manual tasks related to managing organizational computing infrastructure. Automation is nothing new. In fact, it helps to think about the history of automation and how prior efforts transformed the world in far-reaching ways. The first industrial revolution occurred when machines were invented to replace animal and human labor. The invention of steam engines changed the means of production. Later electricity enabled even more automation. Factories changed their approach from using human labor to robotics and machines. Practices were standardized and the craftsman/artisan worker was made largely obsolete, at least for large-scale manufacturing. The same thing has happened in the IT world. At first, IT shops were run by craftspeople who understood the technology and managed to tinker with its capabilities to create amazing (albeit highly customized) systems. These people were treated as IT gods. Organizations could not afford to lose them nor their expertise.

The development of new technologies in computing and networking led to widespread deployment of systems. The only way for vendors to support their products effectively was to develop processes and tools that standardized and automated processes. So, like the industrial revolution years ago, an IT revolution occurred to replace the artisan IT worker with machine-driven processes that would be self-documenting, replicable, and understandable. This also enhanced economic efficiency (Bauer 2018).

Cloud automation is part of this movement. In cloud environments (particularly public cloud environments), the processes and tools an organization uses to reduce manual efforts must be standardized. Cloud vendors need to ensure all clients have access to similar tools and that resources are used in a consistent and standardized way. IT specialists can use cloud automation in any environment: private, public, or hybrid (Karthikeyan 2017).

Cloud Technologies: An Overview of Cloud Computing Technologies for Managers, First Edition. Roger McHaney.
© 2021 John Wiley & Sons Ltd. Published 2021 by John Wiley & Sons Ltd.
Companion website: www.wiley.com/go/mchaney/cloudtechnologies

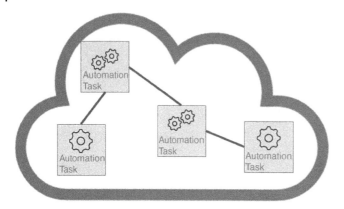

Figure 5.1 Orchestration organizes multiple automation tasks.

Although no cloud deployment is truly 100% automated, vendors continually innovate and roll out new tools. Total automation is a goal!

Orchestration

Orchestration is closely related to automation and sometimes the two terms are used interchangeably but a distinction exists. Automation is a catch-all phrase that describes automating manual tasks. In other words, tasks are deployed without human interaction. Orchestration is a type of automation, but it really means coordination of automation tasks in ways that make sense on multiple levels. It is good to think about the relationship of these terms. Orchestration is a special automation category that organizes multiple automation tasks. Figure 5.1 illustrates.

Terms Related to Automation

Automation: Replacing manual tasks with automated solutions.
Automation Task: A specific element of automation. One item.
Orchestration: Organizing automation tasks to ensure the tasks work together.

Automation tasks can include many items in a cloud environment. For instance, an automation task might be spinning up a Virtual Machine (VM), installing a desktop image, or deploying an application. Automation tasks use tools that might range from a script to a configuration management tool provided by a cloud software vendor. Orchestration, on the other hand, has a goal of end-to-end task workflow automation. Or, at least it seeks to coordinate and organize the processes that an IT team has put into place. In a sense, orchestration is a service that may have to operate at several layers. So, in general, automation and orchestration have a common goal: reduce manual IT management tasks.

Automation Tasks

Cloud computing tasks have a wide range of possibilities when it comes to automation. While the goal of 100% automation might be far-reaching in current environments, many tasks lend themselves well to automation. Having "robots" monitoring systems and dealing with mundane tasks can save resources and make operations more efficient. Example automation tasks can include provisioning or decommissioning servers to meet current resource needs; backing up data and systems regularly; configuring IT resources; creating replication data sets for analytics or web server usage; and, zombie data or cloud instance detection.

Example: Blade Automation

Many IT shops rely on blade server systems. This means adding a stripped down, high-performance server computer into a Cisco Blade Server Chassis for more Central Processing Unit (CPU) power. Modern automation hardware and software combinations do the work. Automation tasks such as this save human interaction. For instance, data center personnel may not have time to configure, deploy, or remove blades in an ongoing fashion. Scripts, hardware solutions, and templates enable administrators to control operations from a console. Interaction is reduced and efficiency is enhanced. Consistency also is enabled.

Any manual operation done in an IT shop is more likely to include elements of human error. Think about completing repetitive tasks like sizing, provisioning, and configuring a VM. Think of creating logical unit numbers for storage or other operations that could involve typing in and tracking unique combinations of digits and then monitoring these to track and assess performance. IT professionals have been doing these sorts of tasks for decades and often are very good at their jobs. However, manual operations can result in errors, and these errors may be difficult to discover. Errors also can result in security issues and open organizational resources to potential risk.

Cloud automation can enhance IT effectiveness and reduce error. This saves time and money. Automation cannot always stand on its own but can be part of a broader vision of integrated automation – orchestration. Often, this means IT administrators take steps to integrate efforts in a strategic manner. Orchestration codifies, integrates, and controls workflows of automation. Figures 5.2 and 5.3 illustrate how an IT shop might start with automating tasks and then move to orchestration.

Implementing Orchestration with IaC

The "holy grail" of automation and orchestration would be a completely self-managed cloud environment that intuitively anticipates resource needs and uses AI to keep costs low, utilization high, and all user needs balanced. Until that is accomplished (which is in the foreseeable future), several use cases exist for cloud orchestration. One of these is to implement an approach known as *Infrastructure as Code (IaC)*. IaC is like using program scripts to automate mundane tasks but takes this practice to the next level.

In standard program scripts, steps for task completion are put into place. These steps are reusable, but only with modification. In many IT shops, scripts are stored in GitHub or other repositories and help organize best practices and prevent the operations team from reinventing the way to bring up servers, perform backups, configure desktops, and many other tasks. In less organized shops, unique scripts for the same tasks might be written by different team members. This practice can result in inconsistent configurations and numerous other

Figure 5.2 Standalone automation tasks in a typical IT shop.

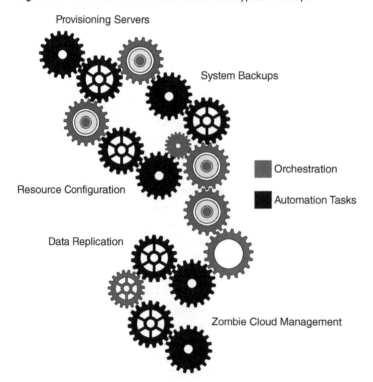

Figure 5.3 Orchestration ties standalone automation tasks together.

Figure 5.4 Example IaC script.

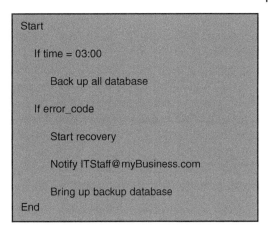

```
Start

  If time = 03:00

    Back up all database

  If error_code

    Start recovery

    Notify ITStaff@myBusiness.com

    Bring up backup database
End
```

problems. These scripts may have been well-written and perform as expected but slight implementation differences would exist. Likewise, documentation could vary or be omitted by some team members.

IaC, on the other hand, removes these issues and uses higher level programming languages to adaptively modify and reuse code across servers and locations. For example, an IaC developer will write a procedure for performing a task. Maybe the task is for setting up an SQL database with a set of tables initially loaded with a dataset. The procedure also sets up User IDs and passwords. The IaC developer approaches the development effort as any software engineer might: following quality assurance guidelines, using version control, testing, and so forth. A systematic and organized approach to IT operations is ensured. It is done once and used by all team members.

Various IaC tools are available to help with automation and orchestration. In general, three approaches are used: declarative, imperative, and intelligent. Declarative tools focus on the end configuration or the "what." In other words, they describe the desired state and let the system execute steps needed to reach that state. SQL is an example of a declarative tool. In SQL, the desired outcome is specified (e.g. what data should be retrieved but the steps of doing that are left up to the system). Imperative approaches describe the "how." In other words, what procedural steps are needed to reach a desired end-state. A programming language such as C# or Java can be used this way. The final approach, intelligent, focuses on the "why." Why should the configured system operate a certain way? This approach seeks best practices and implements automation in ways that mirror these practices. So, knowing the co-relationships of multiple applications on the same infrastructure provides information for this approach. Figure 5.4 provides an example of how IaC scripting looks.

IaC Example

So far, we have looked at IaC from a high level. A concrete example makes more sense. Imagine for a moment that you work for an organization that provides online websites that store information in a database for its clients. In general, each client needs a web server and database backend. Each time a new client is acquired, you log into your Microsoft Azure platform, create a database VM and a webserver VM. You provide a user ID and password, set up basic security and other items for the new cloud infrastructure. So, each time you get a new client, you must go through the same steps. Maybe multiple people work in your organization. Most of the employees use similar, but not exact, steps to create the two VMs, webserver, database, and so forth for new clients. Sometimes, one of the employees forgets part of the procedure and someone must go back and fix the infrastructure. Now, imagine that instead of reinventing the wheel, you have a tool that records how you configured your base "client system." Instead of manually going through the process each time, you use the tool that runs the code and each time a correctly configured client system is deployed. That code is maintained in your IaC tool and allows you to automate, test, and reuse your infrastructure code. You also control the version of the code and ensure your best practices are retained.

This was a very simple example meant to illustrate how the process works. IaC can be useful in many areas of system administration. Among the common uses are:

- *Resource Deployment*: Setting up new resources as needed.
- *Configuration Management*: Configuring new resource instances in complex environments.
- *Updates*: Deploying changes to multiple servers.
- *Provisioning*: Changing resource capacity of VMs as needed by ongoing demand.
- *Load Balancing*: Automatically balancing server loads in ways that ensure optimal system usage.
- *Security Updates*: Applying security patches across various resources.
- *Problem Identification*: Monitoring ongoing activity and pre-emptively finding and correcting problems.
- *Alerts*: Sending out alerts based on automated monitoring.
- *Collecting Information and Monitoring*: Tracking key metrics about system use and performance from multiple machines or VMs in various data centers.
- *Backups*: Ensuring backups occur according to a schedule and automating related tasks.
- *Rollback*: Taking servers back to configurations prior to a selected date or time.
- *Disaster Recovery*: Ensure production servers are continually synced with off-site backups in case of disaster.

Idempotence

Benjamin Peirce, a Harvard mathematician in the nineteenth century, developed the term *idempotent*. This term originally described algebraic elements that did not change when raised to a positive integer power. In cloud computing, it describes the goal of IaC: ensuring code produces the exact outcome each time it is used. This contrasts with people performing tasks manually in ways that could have slightly different results. So, IaC ensures resources are idempotent. Figure 5.5 shows Professor Peirce.[1]

---

1 Unknown photographer (http://www.pragmaticism.net/faq.htm) [Public domain], via Wikimedia Commons.

Figure 5.5   Professor Peirce.

IaC Tools

Many IaC tools have emerged to help DevOps specialists automate their operations. In general, IaC tools configure and automate infrastructure-related tasks, particularly mundane, time-consuming ones. These tools provide task automation and orchestrate deployment and integration of tasks. They usually include roll back features which allow infrastructure changes to be undone should unexpected problems arise due to the update. IaC tools help stop *environment drift* in the resource release pipeline. In other words, IaC stops IT specialists from sprinkling in custom features. Without IaC, over time, each environment becomes a snowflake. That is, a unique configuration that cannot be reproduced automatically. Inconsistency among environments leads to issues during deployments. With snowflakes, administration and maintenance of infrastructure involves manual processes which are hard to track and ultimately contribute to errors. Most modern IaC tools offer configuration and monitoring options as well as work with templates using either a "push" or "pull" approach.

Push Approach
This method uses a centralized server to push out infrastructure changes. It allows administrators to carefully control production environment changes and deployments.

Pull Approach
Client resources contact a central server to request updates. The central administrator has less control over configuration of resources. This approach often is used in places where client users have the expertise to manage their own systems and need to manage their current configuration. A program development shop might be a good example of where this approach would be used.

   IaC tools generally fall into two vendor categories: third party supplied, and native cloud provided. Example third party vendors include Puppet, Chef, Red Hat Ansible, SaltStack, and Terraform. Native cloud tools include Amazon's AWS CloudFormation, Google's Cloud Deployment and Microsoft's Azure Resource Manager. Some tools include scripting languages unique to their products. Others use standard scripting formats like JSON. Most include a template that can be filled out with declarative statements and used to drive deployment in a consistent and reproducible manner. Details on products from both categories follow, starting with a few third-party vendor tools.

Puppet

This tool started as an open-source configuration management tool running on Unix and Windows systems. Like CloudFormation, it uses a declarative language to describe desired system configuration. Originally, Puppet managed configuration on individual physical computers. It then became available for virtual and cloud-based systems. An enterprise version of the software, Puppet Enterprise, provides tools to manage configurations and support migration from physical data centers to cloud-based solutions. Puppet uses files called *Puppet manifests* which are like templates. The manifests are created with Puppet's declarative language or Ruby commands. Puppet has a utility program (Facter) that discovers existing system information and creates a catalog with all resources and dependencies (Hall 2016).

Chef

This automation framework is widely used to setup, configure, deploy, update, manage, and remove servers and applications in a range of environments including cloud-based ones. For instance, Facebook uses Chef to manage its many thousands of servers. IT infrastructure is described in recipes which are used to set up infrastructure. The idea behind recipes is that each time a resource is provisioned, it is identical to all others. In a sense, a cookbook holds the details of an entire infrastructure. This resembles a cookbook used in a restaurant to define all the items served on a menu. So, if disaster were to occur, an enterprise's IT infrastructure could be rebuilt in a new cloud very quickly using the recipes.

Chef Concepts	
recipes	Recipes are at the heart of Chef. They define resource configurations and are used to install, update, and deploy IT infrastructure elements.
cookbook	Cookbooks contain many recipes and define scenarios to automate installation and deployment of resources. Think of cookbooks as orchestration of recipes.
workstation	Station where Chef recipes are developed and deployed to the Chef server for use by the nodes.
Chef-server	Central controller for the Chef system that serves out recipes to nodes.
node	Any physical, virtual, or cloud machine served by the Chef system.
Chef-client	Agent running on every node to assist with Chef-server interaction.
Chef Infra	An automation platform used to transform infrastructure into code.

Chef can help with task automation, for instance determining that all client web servers have access to separate SQLServer instances. Or Chef can help with orchestration by creating an entire infrastructure which uses recipes in an integrated way. Chef comprises three primary elements: workstations, Chef server, and nodes. An IT specialist develops and tests code on the workstation. The perfected code is deployed to the Chef server which makes it available to the nodes. The nodes are client machines in the infrastructure. The client machines download and run the code to update their configuration.

SaltStack

This platform, often called Salt, was developed by Thomas S. Hatch to manage large scale infrastructure when he became frustrated with existing programs (such as Puppet and Chef). The result was a Python-based open-source software system intended to support the IaC approach to enterprise IT resource management. Originally, SaltStack only worked on Amazon Web Services (AWS) and that has been its primary focus although recent implementations have been developed for other environments such as Azure and Google Clouds. SaltStack is considered by many experts as the most scalable, best system for very large networks of servers. It has highly customizable dashboards and consoles to enable IT staff members to monitor performance and anticipate problems.

Terraform

This is another IaC tool used to create, change, and update IT resources in cloud environments. It uses a series of code-based configuration files to drive infrastructure deployment. It uses a datacenter paradigm to help structure cloud-based environments. Like Puppet and Chef, Terraform uses a declarative approach to defining IT infrastructure.

Cloud Provider Resource Management

In addition to third party tools such as Puppet, Chef, and SaltStack which can be implemented in cloud-based or physical environments, the large cloud providers (e.g. Amazon, Microsoft, and Google) offer cloud resource deployment and IaC tools for their environments. Amazon offers a tool called CloudFormation which was developed for use on AWS. Both Google and Microsoft offer IaC solutions for their environments in addition to supporting available third-party tools. We will look closer at several native cloud management tools.

AWS CloudFormation

This is a configuration orchestration tool developed and released by Amazon that permits IT specialists to codify infrastructure and automate cloud resource deployments. CloudFormation relies on templates which use Yet Another Markup Language (YAML), JSON, or can be visually developed in a drag and drop environment called AWS CloudFormation Designer. CloudFormation is declarative. As mentioned earlier, this means the desired environment is described and the tool transparently creates the steps to achieve that outcome. This also means that the order of creating and deploying resources is managed by the tool, although a manual override does exist. CloudFormation's template is a JSON file. The template defines the desired application components. The template is submitted to a server which builds a running instance of the template. The running instance is a "stack" which can be edited after deployment and managed with a version control system.

What Is YAML?

YAML, which rhymes with camel, is a markup language with data-oriented use and intention. The acronym YAML originally meant, "Yet Another Markup Language." Later, the developers suggested its name was recursive and really stood for: YAML Ain't Markup Language. To best define the use of YAML, the inventors, Clark Evans, Ingydöt Net, and Oren Ben-Kiki provide these YAML design goals in decreasing priority[2]:

1. YAML is easily readable by humans.
2. YAML data is portable between programming languages.
3. YAML matches the native data structures of agile languages.
4. YAML has a consistent model to support generic tools.
5. YAML supports one-pass processing.
6. YAML is expressive and extensible.
7. YAML is easy to implement and use.

Google Cloud Deployment Manager

Like CloudFormation, Deployment Manager facilitates IT shop needs for declaring, configuring, deploying, and managing cloud resources. It uses JSON to describe deployment rules and supports YAML as well.[2] A valuable use case for Google's Deployment Manager occurs when numerous, repeated configurations need to be provisioned. This could help with identical development, testing, and production environments or when a resource is used repeatedly for many different users. Once a resource or environment is templated, it can be recreated very quickly without needing a team to research and reinvent what was done in the past. Likewise, when a resource is already in existence, the template can be used as documentation. Minor changes or updates are implemented easily within a template then deployed for each instance. As with third party tools, Google encourages IT specialists and infrastructure developers to integrate the template with a source code control system to manage changes, revisions, and versions.

Google suggests a best practice is to start with Deployment Manager templates rather than create infrastructure or resources directly on a Google Cloud instance. Deployment Manager has two components. One is Template and the other is Deployment. A Template is the declared resource or infrastructure elements. For example, a SQL database might include settings, initial table structure, firewall rules, time zone, and other configuration items. A Template can be either a new item or a modified version of an existing one. Deployments are instances of Templates. Think of the Template as a blueprint and the Deployment as the physical (or virtual), running object itself. Each Template is unique whereas many Deployments exist. So, Deployments turn declarations of resources into

---

2 From YAML Ain't Markup Language (YAML™) Version 1.2, 3rd Edition, Patched at 2009–2010-2101 by Oren Ben-Kiki, Clark Evans, and Ingy döt Net at HTML: http://yaml.org/spec/1.2/spec.html. According to their Website, Copyright © 2001–2009. This document may be freely copied, provided it is not modified.

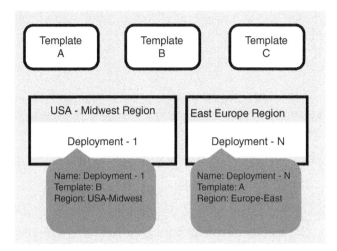

Figure 5.6 Templates are used to create Deployments with Google's cloud Deployment Manager.

actual running and usable artifacts. Figure 5.6 shows how Templates can be used as blueprints for various deployments.

Azure Resource Manager

Microsoft offers a native infrastructure management tool for Azure cloud. The tool considers the complex environment that Azure may be asked to deliver replete with VMs, networks, storage, databases, webservers, microservices, third-party software installations, and so forth. Microsoft developed considerable expertise over the years building tools for system administrators to create images deployed over a network to create desktops for large numbers of corporate office workers. Azure cloud permits that activity plus it extends the management and resource deployment into underlying infrastructure. Since resource deployment may differ depending on user requirements, Azure Resource Manager provides facilities to develop and work with differing resource groups. In a development environment, this might be coding, testing, staging, or production. Each group may require similar resources (e.g. databases, VMs, networks, and so forth) but different security, auditing, and user features. Unique templates address different needs and still maintain continuity. Azure Resource Manager, like other tools, analyzes resource dependencies to ensure deployment occurs in the required order. So, for example, if a Web Server needs to use a database to operate, a dependency is set. Azure Resource Manager introduced several new terms to describe its capabilities and functions. Among these are the following.

Resources

Any item available through Azure. Example resources are VMs, virtual networks, Web Apps, databases, web servers, blobs, storage networks, and so forth.

Resource Groups

In a sense, resource groups represent orchestration. Containers hold related Azure resources for each group or set of Azure solutions. A group may represent a user role, an organizational use, or a complete configuration for an organization. A group will hold all resources for a solution and allow the IT specialist to manage those at a single point. Business rules and organizational needs generally drive the content of resource groups.

Resource Providers

Azure Resource Manager provides a service that helps IT specialists determine what resources can be managed within the software. Each resource provider that is available and that provides options for working within Resource Manager is listed. For example, virtual machine resources are provided through Microsoft.Compute. Microsoft.Web provides details related to web apps.

Resource Manager Templates

This tool provides access to the JSON template files that describe how resources are deployed for a group or resource. The templates are created using declarative syntax. The sequence of deployment is not specified. The deployment outcomes are described instead.

Template Deployment

The deployment options in Resource Manager allow an IT specialist to determine which template to deploy. The deployment uses the template as a blueprint for an instantiated instance of the resource on Azure.

Microsoft offers best practices for Resource Manager and its capabilities. Specifically, they suggest all resources in each group should be synced to share the same lifecycle. Since groups deploy together, their timing must match. If a resource does not match the others, it should be moved into a separate group. Each resource can only belong to a single resource group. This does not mean, for instance, that all SQLServer DBs must be in the same group. It might mean that different installations of SQLServer are treated as different resources. On the same note, resources can be removed or added to groups easily at the time needed. An entire organizational IT infrastructure does not need to be deployed with a single template. Instead, it is best to break it up in ways that make sense considering how the organization operates. It is possible to create templates and then link them with a master template that pushes out the changes needed at that point in time. Figure 5.7 illustrates the idea of a master template and Figure 5.8 provides an example of how template code might appear.

Access Control for Resource Management Tools

Nearly every resource management tool used for automation and/or orchestration has controls to ensure access according to business rules. Essentially, access control (which we will cover in more detail later in this book) allows a business to determine which people and groups of people should have access to specific actions regarding various resources. Most tools integrate some form of role-based access into the management platform to ensure resource deployment, changes, updates, provisioning, and deprovisioning are done by the correct owners.

Role-based access in resource management software is based on two primary concepts: *role definitions* and *role assignments*. Role definitions describe a permission set. This set is given a name and is used in multiple assignments. The definitions can be hierarchical and

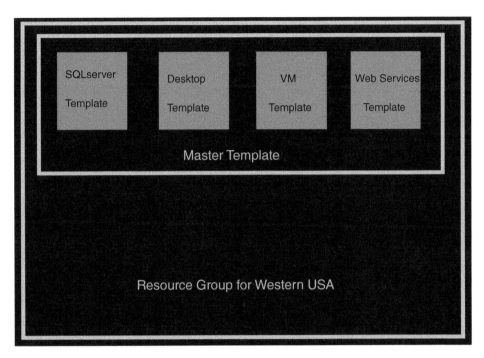

Figure 5.7 Nested templates are often used with Azure Resource Manager.

```
"resources":
[
  {
    "apiVersion": "2021-06-15",
    "type": "Microsoft.Compute/VMAccounts",
    "name": "VMAccountWUSA",
    "location": "westernusa",
    "sku":
    {
      "name": "Standard_VM"
    },
    "kind": "VM","properties":
    {
    }
  }
]
```

Figure 5.8 Example JSON script for Azure resource manager.

inherit definition sets from other definitions. Role assignments attach a definition to a user or user group. Users are generally assigned to predefined roles. The roles may offer different users or individuals, various levels of access and control over resources. Some tools allow setting programmatic access levels to enable scripts to access resources, so an individual's access level does not need to be used. And, many resource management tools offer *locking* features that prevent the accidental removal of key resources without taking extraordinary measures.

Customized Policies

Most resource manager software used for automation and orchestra provides the ability for IT specialists to create customized resource management policies. Policies cover many areas including resource usage, scope of governance, and naming conventions. Often, policies relate to security and ensure checks and balances are in place for user role assignments. Naming conventions may include specific conventions that consider region, resource type, user group, and other items. Policies may also require that resources be tagged with standard identifiers that make it easier to perform searches and logically track usage. Policies may also specify the business owners of various resources.

Default Roles in Azure

Azure Resource Manager provides several default roles regarding resource access and management. These can be grouped into two general categories: platform roles and resource-type roles. The following lists summarize examples of each.

Platform Roles

1. Owner – manages all resources and able to assign access to others.
2. Contributor – manages all resources but may not assign access to others.
3. Reader – cannot change resources but can see everything.
4. User Access Administrator – Can manage resource access settings.

Resource-Type Roles

1. VM Contributor – manages VMs but may not assign access to others.
2. Network Contributor – manages virtual network resources but may not assign access to others.
3. Storage Account Contributor – manages storage accounts but may not assign access to others.
4. Website Contributor – manages websites but may not administrate plans or authorize users.

Microsoft provides more Azure role information at these two sites:

https://docs.microsoft.com/en-us/azure/role-based-access-control/built-in-roles
https://docs.microsoft.com/en-us/azure/role-based-access-control/role-assignments-portal

---

APIs and SDKs

Cloud management often relies on the use of *Application Programming Interface (API)s* and *Software Development Kit (SDK)s* for implementing software or tools on platforms. Many of these are open source and have been adapted to permit use of a package in a cloud environment. We will look a little closer at both APIs and SDKs in the next two sections.

APIs

An *API* is an interface that allows one piece of software to talk to another. At its essence, it is a translator that takes the conventions of one system and transforms it to the conventions of another. Imagine for a moment, a Finnish person attempting to communicate to a Dutch person when neither understands the others' language. A language translator who understands both can provide the link between them. That resembles the role of an API.

Many types of APIs exist and even software intended to work together seamlessly may use an API as an interface because the technology ensures consistency. In general, APIs convert commands from one resource into a usable command on another resource. APIs not only exchange data, they may also perform validation operations on the requests. In cloud environments, APIs permit software systems to communicate and interact. In fact, there are multiple ways that APIs are used in cloud infrastructure. The following sections illustrate.

SaaS APIs

The best-known use of APIs has been traditionally in the application layer. End-user software such as Enterprise Resource Planning (ERP) systems can be connected using APIs. In a sense, the API extends the software package into capabilities used on the cloud. A good example is Microsoft's Power BI. It has a desktop version but many of its capabilities, including sharing visualizations and reports are enhanced with its cloud-based extension.

PaaS APIs

Platform-based APIs provide integration with database services, messaging systems, communication interfaces, and other areas.

IaaS APIs

These are relevant to automation and orchestration services and help control cloud resources and implementations. APIs can control resource distribution and interaction at a high level. Additionally, APIs can integrate and link multiple cloud providers or enable different platforms to interact.

Several different API providers supply software and services to cloud vendors and users. Among the market leaders in API-based services are:

- *Apache CloudStack*
- *Amazon Web Services API*
- *Google Compute Engine*
- *OpenStack API (which we will cover in more detail later in this book)*
- *VMware vCloud API*

Although each API provider offers slightly different services with its own focus, all seek to provide consistency and interoperability between different platforms.

> **Web APIs**
>
> These have become popular for allowing data from different systems to be accessed and shared via the Internet. Many social media systems permit extraction of data based on APIs that limit users to those with an assigned API key. Other systems, for example Twitter, use an API for nearly all user operations. Web APIs really are instruction sets that permit users from the web to access and interact with a system according to pre-defined use cases.

SDKs

Software Development Kits or *SDKs* permit IT specialists to use a software package or resource on their platform. SDKs generally include tools, samples, pre-written code, processes, documentation, guides, and other resources to make implementation easy. While an API is a standard way to access a resource, the SDK facilitates recreation of the resource in a new space. Many SDKs are open source and provided without charge through GitHub. Others are created by the software developer and they may charge a fee or build the cost into their product.

SDKs intended to facilitate cloud deployments have been developed for many platforms. For example, Azure has open source SDK repositories for .NET and for other software resources such as Java, PHP, Python, Ruby, Xamarin, and Android.

SDKs and APIs

It is not uncommon for SDKs to include APIs as part of their toolset. This means if someone creates new software that needs to "talk" to a web service or other resource, an API may be used for the communication piece. An SDK contains the building blocks for an application and an API is a very important block in the set. SDKs usually include APIs, but APIs do not normally contain SDKs. SDKs are meant to aid in application creation while APIs enable communication between resources.

Cloud Backup and Replication

Cloud backup and replication are important management activities that emphasize how cloud technologies can remove the burden from on-premise IT Staff. Cloud backup, of course, stores resources at points in time so systems can be returned to prior states should the need arise. Backups can occur very frequently (e.g. every update is backed up) or less frequently if systems rarely change or data can be easily recovered. Replication, on the other hand, is more about data protection, integrity, and transaction speeds. Replication may occur because of application migration processes, data warehousing, data mining, and analytics operations. Data replication enables more uses of existing data without interfering with critical operations and helps ensure data security. While both backup and replication use similar approaches, many differences exist. Both processes benefit from the cloud's economies of scale and elasticity, and can benefit from automation and orchestration capabilities. We will cover more details in the next sections.

Cloud Backup

Cloud backup is insurance against events that can prevent a business from maintaining *business continuity* (*BC*). BC means ensuring operations remain as normal as possible under all conditions. For example: catastrophic weather events – such as a tornado or hurricane, equipment failures, cyber security failings, or other terrible things. Since IT infrastructure and resources are key elements of BC, it makes sense to backup these elements. In the past, this usually meant having a legacy, offsite backup, and recovery system. Production data was stored on tape or other media and kept in a secure location. Usually, this was a different geographic location. Often, key equipment was kept ready to fill in should physical infrastructure need rapid replacement. Recovery processes usually took hours, days, or weeks to get systems back to a useable state. Today, a much better option exists in the form of cloud backups.

Most vendors of cloud backup systems market two types of solutions: (i) backup/recovery of on-premise resources to the cloud; and, (ii) backup/recovery of cloud-based resources and systems to cloud-based storage. Most backup systems rely on point-in-time triggers driven by business policies. Costs of backup versus the potential costs of loss/recovery time are considered in formulating these policies. Additionally, policies govern how long to keep backups (which may be influenced by government regulations and laws), how many backup copies to keep, and the form of backup (e.g. should everything be backed up each time or only changes since the last backup). If mission critical systems are in place, should they be backed up continually? Another consideration is whether compression software should be used. If so, how does this impact recovery speeds?

Other considerations involve the methods used to backup on-premise data. Does an end-user select the data to be backed up by installing a network drive which is backed up by the IT staff? Or are all on-premise data protected? As it seems, data and system backups are very complex issues. And, if not properly implemented, can result in extreme ramifications both to an organization and an IT specialist's career.

Business rules and organizational needs should govern backup policies. These policies are as varied as businesses in general. To provide an example, consider the following BC policy snippet (most BC policies cover more than backups, but that is our focus here). The backup portion of the policy may specify every Sunday evening a full backup is performed using pre-selected compression technology. On the other six days of the week, incremental backups occur to ensure all new material is stored. The backed-up material is stored in the cloud where it is off-site and safe. Most cloud providers have their own backup and redundancy built in as additional safety. The backups are maintained for at least six months before being removed. Finally, a year end full backup is stored and put into an archival system.

Organizations having multiple locations with on-premise servers or specialty systems would be particularly well-served to have a cloud-based backup for protection. Each location could have its own policies and then use central cloud storage to restore any data lost

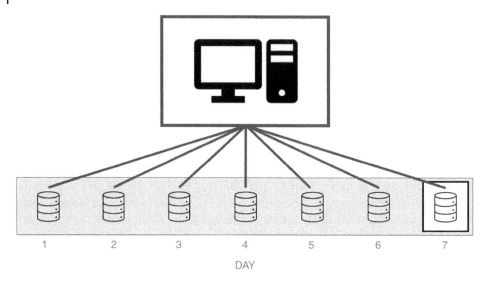

Figure 5.9 Example backup policy that specifies partial backups for six days, then a full back up on day seven. Source: Based on Dracos, 2011.

during a catastrophe. So, copying on-premise data to the cloud leverages economies of scale and helps implement BC policies. See Figure 5.9.

A second form of cloud storage is becoming more widespread. Instead of the local organization becoming involved with formulating and implementing a backup policy, the cloud vendor takes on this role.

Cloud Backup Processes

Cloud backup services for organizations rely on software solutions that facilitate managing the complexities of reliability and rapid restoration speeds. The cloud provides elasticity as a key asset related to cloud backup systems that evolved from Network Attached Storage (NAS) and other traditional storage systems. Newer systems manage backup in sophisticated ways that enable finely granulated automation. This means backup procedures can manage individual devices on corporate networks or specific data objects. The results are custom, flexible, hierarchical storage plans suited to specific business practices.

From a cloud vendor perspective, backup services make great sense. For them, it means a large pool of storage is created and then used to serve the needs of their business clients. The storage pool is managed to ensure it is safely stored in multiple locations with safeguards in place to prevent data loss. The cloud clients pay for peace of mind, knowing a vendor will protect their data. Cloud providers can offer tiered pricing depending on customer's needs for security, retention length, government regulations, speed, and storage sizes. Some cloud vendors may also install on-premise NAS devices for their clients to ensure high-speed storage capability and low latency for recovery. Several options exist and can usually be tailored to customer needs and budgets.

Vendors also provide best practices for their client that may not have storage expertise. For instance, a common practice is known as the 3-2-1 rule. This rule suggests that an organization:

✓ Should always maintain three copies of its data.
✓ Should back up its data on at least two different forms of storage.
✓ Always keeps at least one copy of its data offsite.

It is important to use cloud backup services that go down to the business device level. This means mobile devices, PCs, Macs, or any specialty devices used within an organization. As internal IT infrastructure becomes more virtualized, this task becomes more manageable. Many cloud backup vendors have released new tools and software systems specifically aimed at storing data from mobile devices. This is particularly important in IT environments that are dispersed or that have geographically roaming team members.

Cloud Backup Drawbacks

The idea behind cloud backup is sound but it does come with some implementation considerations and downsides. First, if an organizational IT infrastructure is located on-premise, backing up to the cloud can be slow and tend to take over bandwidth and resources. In an SME that operates in a single geographic region and closes its office doors every day at 6 p.m., this might work fine because backup can take place after hours without causing a slowdown. In organizations that operate 24/7, back up times can become a dilemma. Likewise, a challenge exists for people using laptops or mobile devices when trying to complete a backup on a regular schedule, particularly if a large amount of data must be synched – or, if these devices power down when not in use. Solutions do exist for these issues, for instance adding an on-premise NAS device with more sophisticated software to perform local, targeted backups that later are synched with cloud storage.

Another issue is recovery times. If an on-premise system must be restored from cloud resources, assuming working hardware is available and configured, downloading all the data may take considerable time and effort. Some storage vendors work around this by express delivery of physical storage devices preloaded with the data. This could still require a full day to get the hardware to the implementation sites.

Other concerns with backup include custody of sensitive data. Most cloud vendors offer encryption services. Having a secure backup in this instance is a must. News media and customers are focused on recent data breaches and the exposure of their data to malicious hackers or unsavory individuals. Currently, most backup products utilize Secure Sockets Layer (SSL) encryption to protect data as it travels from source to cloud backup location. Once in the cloud, the vendor should provide Advanced Encryption Standard (AES) protection. Many other sophisticated approaches exist. We will examine several of these in a future chapter of this book.

SSL and AES

SSL and AES are two forms of security for data. Both are important when considering cloud backup processes but for different reasons. First, SSL, focuses on encryption during transmission. Essentially this ensures data cannot be deciphered if a wire is tapped or if data is intercepted wirelessly. SSL relies on a technology where the sender and receiver have exchanged a unique session key required to decrypt the data. AES, is technology developed for securing secret government documents. Often, storage vendors use this technology so neither they nor any intruders can access client data. This eliminates the possibility of insider as well as outside hackers. AES is not meant for use during transmission. The encryption keys are in possession of the data owner and are never transmitted. The data user must have an encryption key to decrypt and use the data. AES security can be used by the client before sending data to the cloud vendor or it can be done on the vendor side. Of course, if a client encrypts their own data, chances of compromise are reduced even further. If the cloud storage vendor never possesses the encryption key, the data always remains fully secure. This stops the possibility that an employee of the cloud storage company will steal the encryption key and access client data.

In organizations with virtualized infrastructure, the cloud backup vendor needs to support restoration that facilitates getting the virtual environment back online quickly. Various vendors offer different levels of capability but often it depends on the IT infrastructure backed up.

Cloud Backup Vendors

Backup is a complex, specialty area and involves more than just copying data into the cloud. Several issues exist to complicate how data can be backed up and recovered. For instance, specialty business applications such as ERP systems, Customer Relationship Management (CRM), Data Analytics, SQL Service, Microsoft Exchange, and others have complex architectures that require specialized data recovery considerations. Most of these offer native backup and recovery support which must be integrated with broader cloud backup procedures used by an organization.

Other considerations include toolsets available for backup and recovery. Several commercial and open source software options exist. Even large vendors such as Google, Microsoft, and Amazon provide specialized open source (and commercial) tools to aid in backup and recovery.

Most backup vendors offer cloud and on-premise versions of their services. For some organizations, a mixed blend of backups may be the best approach. For instance, an organization may use a cloud-based email system with a vendor that provides backup services as part of the overall package. This will not require an on-premise backup. But the same company may be using an on-premise CAD system that does use a local backup system. This data may need to be backed up into the cloud using a unique solution. Many complex combination approaches exist for organizations and this requires special expertise. The vendors

Table 5.1 Sample cloud storage vendors.

Vendor	Overview	Comments
Arcserve UDP Cloud Direct	Useful with all major business platforms	A fully featured business solution vendor for recovery, backup, and BC. Developed for Microsoft environments. No bare metal recovery. Protects against ransomware.
Box for Business	Reliable, widely used vendor	Offers intuitive interface with easy access and many features. Supports many environments and offers read-only options and varying levels of access.
Carbonite	Good small business solution	Useful to recover networks or computers that have become disabled. Widely used.
Dropbox Business	Reliable, widely used vendor	Integrates with many software vendors and provides non-intrusive synchronization.
Egnyte	Focuses on solutions utilizing a combination of local and cloud storage	Solutions for secure data from local storage. On-premise servers add security.
iDrive	Robust small business solution	Provides excellent interface and good encryption. Lacks some software support such as Office 365 email.
MSP360	Customizable backup provider	Writes images to Amazon Cloud or Azure. Lacks some management features found in competing packages.
SpiderOak	Collaboration tool with online hosting and cloud storage services	"Zero knowledge" privacy environment – client is the only one who can view all stored data.
Tresorit	Secure cloud-storage with 2-party authentication	"Zero knowledge" privacy environment – client is the only one who can view all stored data.

operating in this space are varied and segmented according to services. Several vendors that provide cloud storage to small businesses or individuals are listed in Table 5.1.

Cloud backup remains an important and widely used technology, but the cloud offers more options that can align better with BC and disaster recovery. The next section investigates how cloud replication can fill this role.

Cloud Replication

Cloud replication is generally part of IaaS functionality. Where backup helps ensure the long-term safety of organizational data and enables compliance with governmental regulations, it may not be well suited to BC and disaster recovery needs. Traditional backup can be slow to restore and does not always include the infrastructure side of an IT site. For

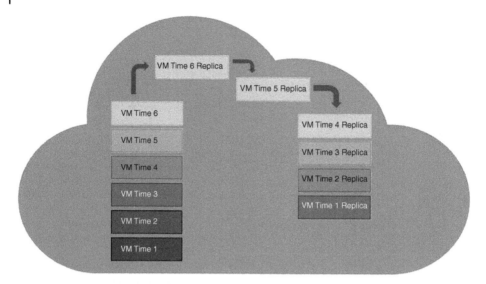

Figure 5.10 Ongoing replication of a cloud-based system seeks to reduce latency and ensure rapid recovery times.

instance, the data might be stored, but software using the data might need to be reinstalled and reconfigured. This could entail licensing issues and become complex on new platforms.

To overcome these sorts of issues, organizations may consider integrating cloud replication into their overall IT solution. Like backup, replication can move data from on-premise systems to the cloud or replicate cloud-based systems to another cloud location. Key considerations include return on investment (e.g. cost of replication efforts versus potential cost of losses without it), implementation costs, resource needs, reliability, and latency of replication. Latency is particularly important in on-premise to cloud-based replication strategies. See Figure 5.10.

In general terms, cloud replication is designed to protect IT infrastructure and data against a catastrophic event or disaster by continually replicating (this means creating an exact copy) data and/or VMs, servers, and other virtual resources in diverse locations. If a failure occurs, a working version of the system can be started in the new location quickly. The time to restart is termed *system latency*.

For an example, imagine that a business has its primary, production system private cloud set up in Wichita, Kansas. Known for being in the tornado belt, Wichita's primary system would be well-suited to have a replication site in a distant geographic location such as the Rocky Mountains, Michigan's Upper Peninsula, the U.S. East Coast or other place far enough away to ensure the same disaster would not strike simultaneously. If a tornado strikes Wichita and disrupts service, the replicated site takes over quickly to ensure BC.

Replication Technologies

Many vendors provide replication technologies. Some focus on database replication and others provide broader solutions. SharePlex is an example of an organization that provides database replication services. It is vendor agnostic (meaning any database can be used with its services). This is particularly good in environments that host databases from more than one supplier (e.g. SQLServer, Oracle, and SAP HANA all used by one organization). Their toolset can work with cloud-based or on-premise systems. The replication software provides comprehensive solutions that consider recovery time requirements, database sizes, transaction speeds, and other attributes. The software provides monitoring, synchronization, and more. Many of the vendors mentioned in the section of this book called *Cloud Backup Vendors* also provide replication services. Recently, vendors have moved away from providing narrow-scoped solutions and instead work to provide broader, comprehensive solutions suited for cloud environments. The next section looks at these in more detail. We will focus on those comprehensive solutions that many vendors market as DRaaS products.

Qlik Replicate

Qlik Replicate (formerly Attunity) is an example platform that focuses on replication. Its unique approach is to provide any-to-any services that permit users to backup any database system and other resources to any replicated platform. Among their features are:

- Universal solution for any type of source: relational database, file systems, mainframes, PCs, mobile devices, SAP, data warehouse, or Hadoop – to any cloud platforms including Amazon, Azure, and Google.
- Fast transfers of data using Wide Area Network (WAN) optimizations.
- High levels of security with data protected by AES and other integrity checks.
- Monitor and user console to control and observe progress and status of tasks.

DRaaS

A wide array of choices exists for putting a BC plan together, and the cloud added even more options. The bottom line for most organizations is the desire for a cost-effective insurance policy that guarantees business operations will be minimally impacted in the event of a major catastrophe whether that be brought on by nature, caused by human error, enacted by a malcontent, or caused through tomfoolery. With the variety of backup, replication, and other services available, many possibilities emerge. To address the complexity and provide best practices solutions, many vendors now market *DRaaS*. At the heart of these services are information technology solutions.

DRaaS makes sense because like all cloud infrastructure, it is elastic and uses cloud resources. VMs, data, containers, and so forth all have a virtual footprint that can be duplicated and put into motion very quickly. The burden is removed from an on-premise IT group as is the expense of having backup hardware/software waiting in case a problem occurs.

All leading cloud providers have explored the DRaaS space and many offer products in this area. For instance, Microsoft Azure offers BC services through Azure Site Recovery. This service provides backup of key resources and data, reporting services, and a console for management of the process. It is low cost and supports physical and virtual infrastructure. It does require a learning curve and must be administered by in-house IT staff.

Other vendors in this space include Veritas with its Resiliency Platform which focuses on rapid recovery speeds; Quorum onQ Hybrid Cloud, a scalable and full-featured service that uses a physical appliance to aid in rapid recovery; and IBM Resiliency Services which is expensive but highly reliable.

DRaaS comes in many formulations and these vary as greatly as the vendors providing the solutions. While a turnkey, full system recovery with minimal disruption is the ultimate goal, this may not always be possible due to the mix of resources an organization uses and the budget it has available. It makes sense to ensure DRaaS covers mission critical systems and data. At a minimum, the following items should be covered:

✓ Automatically backup mission critical systems, infrastructure, and data.
✓ Rapid recovery from system disruptions, losses, or complete failures with very little user interaction required.
✓ Flexible recovery options should include entire system recovery, or the recovery of a single database, application, or resource.
✓ Reasonable billing requirements.
✓ Elasticity of solution so it can grow or reduce matching the organization's footprint.

What Should DRaaS Include?

A good DRaaS implementation needs to consider a few important aspects. These include:

- What platforms does the DRaaS product work with? Considerations for mission critical systems, data, platforms, and infrastructure are particularly important.
- Can the DRaaS vendor manage all organizational data storage needs?
- What processes are used to back up organizational systems? How often does this occur? Is it disruptive? And is data synchronized to be backed up on an interval basis?
- Is a local backup being created or is all done in the cloud?
- Does the system interact specifically with applications that may be difficult to replicate?
- How long does system restoration take?
- How long has the vendor been in business and who are their clients?
- What type of recovery testing is recommended?
- Will the backed-up version of the system perform as well?
- How long will the vendor host the recovered system and what costs are involved with the hosting?

Summary

Chapter 5 looked at several topics important to managing organizational IT infrastructure utilizing cloud-based management features. New software and hardware solutions permit organizations to automate many routine functions and then link these automations together to orchestrate a reproducible IT architecture and platform. Because of this, cloud automation and orchestration need to work together. Good automation can be integrated into an overall orchestration approach to ensure platform stability and that organizational knowledge is captured and not reliant on one expert individual. One method for ensuring stability is to use an approach called IaC. This means using a script-based approach to implementation of infrastructure to ensure all processes are documented and are idempotent (e.g. can be exactly recreated). Vendors provide toolsets such as Chef and Puppet to make this easier and more manageable by IT specialists.

Having a stable IT infrastructure is only part of IT management concerns. Another is to rapidly recover should a system become compromised or destroyed. Business continuity planning includes disaster recovery elements for IT infrastructure. Different approaches have traditionally been used. Backups of data, software, and application configurations are common but often take longer to reinstall or implement should something bad occur. Other solutions include a replicated system ready to start should it be needed. Replicated systems are updated as the production system runs and usually can be implemented in much shorter time frames. Of course, the goal for disaster recovery is that an alternate system can be immediately started if needed. Many cloud vendors offer DRaaS to ensure nearly seamless operation should an unexpected event occur. This event might be a natural disaster, a terror event, a malicious insider, or any number of other disruptions. DRaaS has only recently become available because of the power and capabilities offered by the cloud.

Our next chapter will continue to explore the Ops side of DevOps and describe additional tools used by IT specialists to ensure smooth operations of organizational IT systems.

References

Bauer, E. (2018). Cloud automation and economic efficiency. *IEEE Cloud Computing* 5 (2): 26–32.

Hall, J. (2016). *Mastering SaltStack*. Birmingham: Packt Publishing Ltd.

Karthikeyan, S.A. (2017). Hybrid cloud automation. In: *Azure Automation Using the ARM Model*, 119–140. Berkeley, CA: Apress.

Further Reading

Amato, F., Moscato, F., and Xhafa, F. (2018). Cloud orchestration with ORCS and OpenStack. In: *International Conference on Emerging Internetworking, Data & Web Technologies*, 944–955. Cham: Springer.

Bahadori, K. and Vardanega, T. (2018). DevOps meets dynamic orchestration. In: *International Workshop on Software Engineering Aspects of Continuous Development and New Paradigms of Software Production and Deployment*, 142–154. Cham: Springer.

Baur, D., Seybold, D., Griesinger, F. et al. (2018). A provider-agnostic approach to multi-cloud orchestration using a constraint language. In: *2018 18th IEEE/ACM International Symposium on Cluster, Cloud and Grid Computing (CCGRID)*, 173–182. IEEE.

Ebert, C., Gallardo, G., Hernantes, J., and Serrano, N. (2016). DevOps. *IEEE Software* 33 (3): 94–100.

Eriksson, M. and Hallberg, V. (2011). *Comparison between JSON and YAML for Data Serialization. The School of Computer Science and Engineering*, 1–25. Royal Institute of Technology.

Hummer, W., Rosenberg, F., Oliveira, F., and Eilam, T. (2013). Testing idempotence for infrastructure as code. In: *ACM/IFIP/USENIX International Conference on Distributed Systems Platforms and Open Distributed Processing*, 368–388. Berlin, Heidelberg: Springer.

Nelson-Smith, S. (2013). *Test-Driven Infrastructure with Chef: Bring Behavior-Driven Development to Infrastructure as Code*. Sebastopol, California: OReilly Media.

Paladi, N., Michalas, A., and Dang, H.V. (2018). Towards secure cloud orchestration for multi-cloud deployments. In: *Proceedings of the 5th Workshop on CrossCloud Infrastructures & Platforms*, 1–5.

Prassanna, J., Pawar, A., and Neelanarayanan, V. (2018). A review of existing cloud automation tools. *Asian Journal of Pharmaceutical and Clinical Research, Special Edition for Advances in Smart Computing and Bioinformatics*: 471–473.

Rahman, A., Partho, A., Meder, D., and Williams, L. (2017). Which factors influence practitioners' usage of build automation tools? In: *2017 IEEE/ACM 3rd International Workshop on Rapid Continuous Software Engineering (RCoSE)*, 20–26. IEEE.

Sinha, V., Doucet, F., Siska, C. et al. (2000). YAML: A tool for hardware design visualization and capture. In: *Proceedings 13th International Symposium on System Synthesis*, 9–14. IEEE.

Wu, V. (2016). *Cisco UCS Cookbook*. Packt Publishing Ltd.

Website Resources

Backup Providers

Arcserve UDP Cloud Direct: https://www.arcserve.com/data-protection-solutions/udp-cloud-direct/

Box for Business: https://www.box.com/resources/dm/ccm/home

Carbonite: https://www.carbonite.com

DropboxBusiness: https://www.dropbox.com/business

Egnyte: https://www.egnyte.com

iDrive: https://www.idrive.com

MSP360: https://www.msp360.com/backup.aspx

SpiderOak: https://spideroak.com/one

Tresorit: https://tresorit.com/business

DRaaS Providers

IBMResiliencyServices: https://www.ibm.com/services/business-continuity

QlikReplicate: https://www.qlik.com/us/products/qlik-replicate

QuorumonQHybridCloud: https://quorum.com

VeritasResiliencyPlatform: https://www.veritas.com/availability/resiliency-platform

IaC Providers

Chef: https://www.chef.io
Puppet: https://puppet.com
SaltStack: https://www.saltstack.com
Terraform: https://www.terraform.io

6

What Are Cloud Business Concerns?

Cloud computing has a complex business model that transforms the way accountants and financial managers view the cost of IT. Rather than invest in huge upfront costs and capital equipment, the cloud moves an organization into a pay-as-you-go model. This works well with cost accounting practices that directly tie use to expense. Several tools exist to support business operations in areas such as monitoring and consoles, service level agreements (SLAs), and subscriptions/billing. This chapter looks at these areas and provides insight into the business side of cloud computing with an emphasis on cost and usage considerations. Figure 6.1 provides a big picture view of cloud management considerations. In Chapter 5 we covered automation and orchestration, knowledge retention (e.g. IaC), and backup/recovery. In this chapter we look at cost considerations and performance/monitoring in more detail.

Monitoring and Console Tools

Monitoring and console tools provide several important functions in cloud computing. First, usage can be tied directly to organizational entities. For accounting purposes, this can be very important. Knowing who is using what and to what degree enables more sophisticated approaches to understanding the costs of production, support, and organizational overhead. It also provides ways to avoid or at least anticipate cost overruns. Monitoring enables IT specialists to ensure resources are operating as desired. Problems, issues with capacity, and other important attributes can be watched either directly or through automated management tools that send out alerts. Monitoring also aids in the security process to ensure the correct resources are only accessed by authorized users. We will cover authorization more in Chapter 7. Finally, console tools can be used to aid with deployment and other IT administrative tasks.

Currently, many vendors entered this space with the goal of providing one tool that offers all cloud management needs. While most cloud vendors such as Microsoft, Amazon, and Google offer console tools that work within their platforms, an increase in hybrid and multi-cloud solutions creates a demand for cross-platform tools.

Most experts agree that no one tool currently exists to cover all cloud computing management needs. There are, however, several excellent tools that provide features aligning with

Cloud Technologies: An Overview of Cloud Computing Technologies for Managers, First Edition. Roger McHaney.
© 2021 John Wiley & Sons Ltd. Published 2021 by John Wiley & Sons Ltd.
Companion website: www.wiley.com/go/mchaney/cloudtechnologies

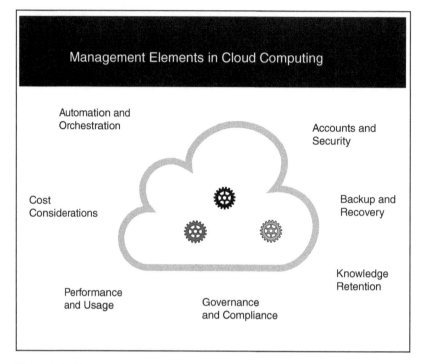

Figure 6.1 Elements related to management of the cloud.

many best practice use cases. It is important for an organization to determine their needs and then match their requirements to a tool that fills most. We explore organizational needs in the following sections, then focus on tools that specialize in these areas. Again, if an organization uses a single cloud platform, in many cases they are better off using the native tools supplied by the vendor. An exception could be if specialized uses need to be considered. If the organization has a multi-cloud platform, then it makes more sense to explore a broader set of third-party tools. In either case, the functions performed are similar. We look at monitoring resources, cost considerations and other topics.

Resource Consumption Monitoring

A primary function of IT administrators in cloud environments is to monitor resources, costs, and consumption. Cloud monitoring uses a set of administrative tools that supervise resources such as virtual machines, servers, databases, and applications to ensure operations continue smoothly within budget and other constraints. Cloud monitoring tools come from two primary sources:

- *Cloud provider tools*: This is a default option because most cloud providers have these built into their platform. No installation is required, and administrators can use these immediately.
- *Third party vendor tools*: These tools can be purchased, often on a subscription basis, to integrate monitoring operations and provide specialized reporting and oversight.

Some IT administrators use a portfolio of tools that include both cloud providers and third-party offerings. In general, these tools seek to proactively detect issues that could prevent IT operations from providing needed services to end-users and may report on performance, speed, resource usage levels, and other metrics. Monitoring alerts administrators to developing problems and aids in troubleshooting. Cloud monitoring provides several benefits to an organization. Among these are:

✓ Costs are contained and controlled.
✓ Interruptions can be preempted and avoided.
✓ Maintenance costs can be reduced.
✓ Paying for more resources than required is avoided.

Planning for Monitoring

Monitoring IT resource consumption and utilization needs to be integrated with overall IT planning. In other words, plans should include metrics key to business success. For instance, if an organization is selling goods online, the customer response times must be kept at an acceptable level. If an organization provides web services, then it must ensure its webservers and databases are not overloaded with requests and that sufficient VMs are provisioned. A monitoring plan should include the following considerations:

- *Identification of key metrics*: An organization needs to recognize its important, key activities and monitor resources that may influence or impact these.
- *Integration of monitoring results*: Reports that summarize key organizational elements should be on a dashboard or other easily accessible medium. Solutions that integrate outputs from different subsystems should be brought together in an easy-to-follow manner.
- *Cost considerations*: Linking costs to resource use should be integrated into reporting. The ability to see where expenses are generated and focus on those areas may be important to many organizations.
- *Identify triggers for action* Particular threshold levels should be identified, and if these are crossed, having automatic notifications sent to key personnel may be helpful. In some instances, actions can be automated. For example, if activity exceeds a level, additional services could be automatically provisioned without human interaction.
- *User metrics*: More than resource usage metrics should be part of a monitoring plan. It is also important to assess user experience. For instance, are response times acceptable, are there any bottlenecks in workflows or resource access?
- *Testing*: Monitoring events during failures and resource outages should be included in the plan. Does the system react as expected? What is the user impact?

Cloud Monitoring Tools

As stated earlier, a wide range of monitoring tools is available for cloud environments. The tool set required for any business needs to consider their monitoring plan and consider budget constraints. An obvious, first-choice selection would be the tools provided by an organization's cloud vendor. If this falls short, then investigating available third-party tools becomes the next step. Popular tools available from cloud vendors include offerings from Microsoft, Amazon, and Google.

Microsoft Azure Monitor: Microsoft offers a suite of monitoring tools to help IT administrators gage the performance of their cloud-based resources. The tool set is built into their Azure cloud infrastructure and using it just requires activating the service. Some monitoring items are free, and others use a subscription model. The software permits highly granular review of resource performance and utilization. It also includes features to enable actions, based on alerts, to be automatically invoked. The Azure portal provides a broad view with drill down features. It also integrates with analytics and monitoring tools. Application Insights is included with the monitor suite. This is a service for oversight used by web developers that need to monitor live web site usage. It provides information on web server operation and provides live data about problems and issues that may impact performance. It integrates with Visual Studio and other Microsoft tools. Its billing structure is based on volume of telemetry data and web tests. Overall, Azure Monitor is an integrated tool that collects and analyzes data from resources running on the cloud. It provides metrics according to use and is loosely structured to match the IaaS, PaaS, and SaaS implementations running. For instance:

- Application metrics describe performance and functionality of software implementations with a specific focus on code developments.
- Azure resource metrics describe the performance and operation of any Azure resource.
- Azure subscription metrics offer details about operation and management of any Azure subscriptions.
- Azure operation metrics describe the health and operation of an organization's Azure implementation.
- Guest OS metrics describe the operation of any virtual operating system
- Container metrics describe the operation and health of any containers implemented in Azure.
- VM metrics describe the operation and health of any VMs implemented in Azure.
- Azure tenant metrics provide information about tenant-level Azure services like Active Directory and so forth.

Azure offers health monitoring to indicate performance and operation status of resources and services. It can provide alerts and notifications based on health. Alerts and other conditions can be used with the Logic Apps tool to automate actions needed to correct problems. Azure offers custom views and dashboards. Diagnostic data is collected and can be used to power custom reports automatically with tools such as Power BI.

Amazon CloudWatch: Like Azure Monitor, Amazon CloudWatch is a native suite of tools provided by a cloud vendor. In this case, it oversees, monitors, and diagnoses issues on AWS Cloud. This system is included and available to all AWS Cloud users. CloudWatch provides data describing resource performance, activity, and usage. It invokes actions automatically to correct ongoing issues or problems. It provides an overall dashboard describing system health and collects operational data in logs, metrics, and events that can be ported into analytics tools for deeper investigation. CloudWatch offers alarms and troubleshooting to ensure applications run smoothly. Amazon CloudWatch uses a pay-as-you-go subscription model for billing.

Google Cloud Operations: In 2014, Google acquired Stackdriver, a provider of cloud management software. It provides performance and diagnostics data derived from operations in Google Cloud. It also operates in AWS, making it a hybrid solution. In general, it offers similar features to Azure and Amazon's native monitoring tools with monitoring, event logs, activity traces, error reporting, and alerts. *Operations* samples application performance data and collects information about how the system is operating. It permits developers to engage with problems and track the source issues more easily. Several visual tools are available to make this easier for developers. This is particularly useful in distributed environments that might have many instances of an application running. Finally, it has features to start and stop code, and fix problems. Operations is more focused on developers than on IT administrators and therefore, may not be the best solution for those running acquired rather than developed resources on the cloud.

In addition to native tools provided by major cloud vendors, a wide range of third-party monitoring tools exist. Table 6.1 provides details of several examples in this area.

Monitoring Challenges

The number and variety of cloud monitoring solutions provide organizations with many options for tracking the health and status of their resources and operations. However, challenges do remain. Included are those relating to hybrid cloud solutions. Not all organizations utilize a single cloud system for all their computing needs. Some may implement part of their infrastructure on one cloud and part on another. Others may use a combination of public and private cloud platforms. And still others may use a combination of on-premise and cloud systems.

Each of these situations may be unique and most experts agree that a single monitoring solution may not be within the reach of all organizations. Many call this a *single pane of glass monitoring solution*. The idea is that an enterprise has one monitor that comprehensively groups all metrics into a single interface and provides drill-down capabilities for that tool. Ideally, the system would be provided by one vendor and perhaps be a layer within the cloud infrastructure. It could become something like monitoring as a service (MaaS).

Other challenges include trying to understand the end-user experience. The best monitoring tools do provide an idea of features needed by users of web apps, microservices and other resources. But, in many cases, the end-user experience is deeply rooted in organizational culture and activities. Therefore, monitoring must be customizable.

The vast amount of data collected by monitoring, tracking, troubleshooting, and logging also provides a challenge. IT administrators seek to pull the meaningful bits out of the data using analytics tools and visualization software. The science and practices behind these ideas are rapidly improving but have not yet matured.

Overall, monitoring systems have come a long way in a short period of time. Major cloud vendors and third-party software providers are working hard to develop solutions that enhance the capabilities of IT administrators and ensure cloud infrastructure operates well.

Cost Monitoring

Monitoring cost is closely related to monitoring the health of IT resources. In many software systems, the two features are intertwined in a way that makes the cost of adding or removing

Table 6.1 Example third party cloud monitoring software.

Software	Overview	Description
AppNeta	SaaS app monitoring tool	Also provides features to monitor networks. Looks at resource usage, user experience.
AppRiver	Monitors web apps, Office 365 and email services	Includes features to protect email systems, reduce spam and encrypt emails.
BMC TrueSight Pulse	Monitors web apps	Runs as an add-on to Amazon CloudWatch or Azure Monitoring and includes visualization features and advanced notifications.
CA Technologies	Full-stack monitoring and cloud management package	This is a fully featured package meant for use by companies running complex cloud solutions. It works with public, private and hybrid cloud systems.
Datadog	Infrastructure monitor	Integrates with many existing cloud software systems to provide additional monitoring features, dashboard capabilities, and visualizations.
Dynatrace	Monitors business analytics, containers, applications, and infrastructure	This is a highly developed mature package that is highly rated. It works with all major cloud platforms and offers a great deal to Google Cloud users.
Exoprise	Works with SaaS applications like Office 365, Skype, Slack, Yammer, Salesforce, and Box	Helps detect problems, outages, and understand the health of organizational software systems.
LogicMonitor	Monitors overall stack of cloud operations	Comprehensive monitoring software that has received numerous accolades and awards.
Rackspace Monitoring	Comes with Openstack cloud accounts and is used for enterprise-wide monitoring	Comprehensive package with excellent notification system.
Retrace by Stackify	Performance monitoring for .Net and Java development	Provides performance metrics, error monitoring, and logs.

resources readily apparent. In general, IT administrators must be very careful of budgets and be aware of trade-offs between capability and added resource costs.

Several factors can increase costs. Of course, adding more VMs, storage, database instances, application licenses, and other more tangible items often is correlated to higher expense. But other subtle issues can impact costs as well. For instance, a surge in web app use might result in higher costs. Or failing to manage deprecated VMs and other resources can result in what are known as *zombie instances*. Monitoring tools can help locate and remove these cost incurring, non-productive items before they affect the bottom line.

Other cost monitoring features can be used to help tie a specific cloud resource to the business owner that uses it. From an accounting perspective, this is helpful and pushes justification back to the resource owner.

Several tools have been developed to specifically aid with cloud cost monitoring and control. We will explore a couple of these to provide a sense of what these tools can do and how they are used by IT Administrators.

Spiceworks Cloud Cost Monitor: This tool is designed to work with AWS or Azure accounts to provide added cost management details. The central feature of this tool is a dashboard with data summaries, trend information, and cost breakdowns that can be drilled into additional details. Spiceworks has an email notification system with the capability to indicate abnormal items. Spiceworks connects with accounts from its own servers and accesses usage data in a read-only manner. Spiceworks offers the following major features:
- Identification and removal of unused or unintentionally provisioned resources. For instance, it may be possible that IP addresses have not been released by partially installed software instances or VMs that are not needed but still occupy service space. This is done from a cost reporting perspective.
- If operational trends show an increase in cost without corresponding increases in demand, the tool can be used to identify why extra bandwidth or other consumption is taking place.
- Regular costs are broken down into more detail both in reports and on a dashboard to help IT administrators track costs and assign these to organizational activities.
- Insightful visualizations of costs over time can be produced.
- Different cloud environments on AWS and Azure can be integrated into one costing dashboard for comprehensive management.

Teevity: This cost management suite works with Azure, AWS, and Google Cloud. Essentially, it is a continuous cost monitoring platform with tools in place to manage and reduce costs. Teevity provides cost information in multiple ways. Some IT administrators prefer to receive periodic emails with cost updates. These might occur daily or be sent as exceptions when anomalies occur. Among Teevity's primary features are a dashboard that provides both current and projected expenses based on historic data and current usage trends. Teevity offers automation tools and APIs to enable action to be taken on cost-related information. Another feature is what Teevity calls its Cloud Asset Inventory. This is a map that lays out all the resources in the cloud in real-time. An auto-discovery function ensures that deprecated instances are identified and can be deprovisioned. The map also gives administrators a high-level view of usage and may help demonstrate areas where resources can be combined. Likewise, resources can be categorized and tagged to provide meaningful and customized groupings for easy review. Teevity also provides specific business costing functionality. For instance, cloud resources can be categorized and allocated to specific business cost centers like projects, departments, or teams; or can be broken down by other factors such as country or region. The classification scheme becomes an element of the dashboard so real time cost values also are visible. Data can be viewed by transaction or by resource or tenant in the cloud. Google's query tools are used to provide the means to create custom information specific to an application. Data also can be ported directly into Google Spreadsheets for further manipulation and use. All billing information can be created and saved to Dropbox automatically so all receipts and required documentation are in one place.

AWS Cost Explorer: This cost management tool is native to AWS and is located within the AWS management console. It provides an overview of costs incurred on AWS in several different formats. First, it provides a monthly report of costs by either services or by linked accounts. It also provides a daily cost report. Like Teevity, AWS Cost Explorer offers forecasting capabilities that estimate three months of bills based on prior activity and current trends. A budget dashboard provides a central focal point for billing and expenditures. Budgets can be created and used to measure ongoing costs. Although only two budgets are included for free, an AWS account can have up to 20,000 budgets for team projects and other activities at a low cost.

CloudCheckr: This tool specifically is designed to enable IT specialists to track cloud-based assets to better understand usage and costs. It outputs a visualization to describe information in a clear manner. Mainly, this tool enables under-used resources to be identified for removal or consolidation to save money. It has built-in capability to report mismatches between resource provisioning and resource usage to help save money. It also provides tools that track security-related costs.

Cloud Cruiser: This is another tool useful in AWS, Azure, and other clouds. It provides usage details and ties these closely to a system of alerts, warnings, and messages to ensure the correct person is notified when an overrun is expected. It can tie budgets to usage and notify managers when budget amounts are likely to be exceeded.

All cost-monitoring tools mentioned can track resource usage, tie that usage to costs, and provide actionable information. Some use alerts and warnings, others provide dashboards, and some visualizations. These tools tie usage and costs back to the business units needing the resources. The marketplace for cost monitoring is currently very competitive and many of the tools mentioned will probably disappear or be bought up by competitors or cloud vendors. In any case, the need for cloud cost specialists and analysts is a niche that will grow and become more important as cloud computing grows.

Costs Associated with Zombie Resource Instances

A concern shared by many IT specialists relates to *zombie instances*. Zombies are the living dead just like on the movies. These are resources that may have once been useful, functioning members of the IT resource pool but are now hanging around, consuming costs without any contribution. Zombies can result from several actions and really, the term, is a catch-all phrase for many problems. First, resources may be created and forgotten. While not a true zombie instance, these are merely unused resources that otherwise are fine. A true zombie instance generally results from an IT-related action that either fails, is interrupted, or is invoked incorrectly. For example, a VM build failure can create a zombie VM that takes up memory space but is not visible to the administrator. A file may be opened and never closed by software that does not manage this function automatically. Again, this is not readily visible, but it takes up memory and may reside in the cloud for a long period of time. For another example, think of your own PC. Have you opened the task manager window and

Figure 6.2 Zombies can cause cost overruns and other cloud problems. Source: Derived from U.S. Air Force photo by 2nd Lt. Siobhan Bennett/Released to public domain.

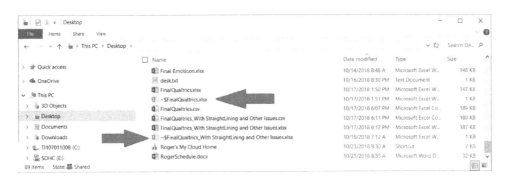

Figure 6.3 Temporary files in Windows take up storage space. Source: Used with permission from Microsoft.

found 20 instances of a program that does not appear to be open, running and slowing your PC's performance? The same thing happens on a larger scale in cloud environments. For this reason, it is important to use a tool that finds and destroys zombies before they take over your cloud and diminish its performance. See Figure 6.2.

Other zombie instances can result from obsolete datasets held in memory. Software that makes updates or changes to files may save a temporary copy and if it is not removed, storage space remains occupied (see Figure 6.3). Consider the impact of the extra storage space in a large organization. Removing even a few files on each employee's desktop could result in sizable savings over time. Further, think of the security ramifications. Having undeleted or temporary files stored in the cloud could be a security concern. On PCs, putting a file in the recycle bin does not free up that space until further action is taken.

Many cloud vendors have built safeguards against zombie issues, but problems remain. Most provide data policies that describe how forgotten data is removed and how long deleted files remain on the system. For file deletions, it is helpful to ensure the vendor securely removes files. Zombie resource instances are a bit different. Unused servers left running eat up costs. Likewise, failed instances might exist, and these too can be expensive over time. Some cloud providers will manage zombies as part of the contract, but most often, those issues fall under the domain of the cloud subscriber and their IT staff.

Software tools exist to help manage zombie-related costs. For example, RightScale is a software solution intended to identify and remove zombie instances. This software specializes in finding:

- Instances stranded in failed booting. This means instances were successfully provisioned and incurred charges but later failed. These may occur from errors in configuration steps or other failures.
- Unresponsive or unused instances. RightScale enables IT specialists to look at dormant or unused instances and remove these. This is an area where special care must be taken because sometimes these instances are part of a replication or failover scheme and need to remain active and in a waiting mode.

Gamification of Management Consoles

TotalCloud is a third-party cloud vendor that has created an immersive tool to control cloud resources. The idea is that gamification provides a natural way to interact with a system and this will motivate users to optimize resource usage. TotalCloud is inspired by Warcraft and provides a visual and aural environment to "move" IT specialists into their cloud infrastructure. The goal becomes fixing security issues, optimizing resource usage, fixing problems, ensuring compliance, or finding unexpected cost savings. What a great idea! Fun at work for sure!

Service Level Agreements (SLAs)

An *SLA* is a binding contact between a cloud vendor and their client. Generally, the cloud vendor is an organization like Microsoft or Amazon, and the client is a business organization moving their IT infrastructure into the cloud. The SLA describes conditions for use and minimum level of service and performance the cloud vendor will provide. Among the service and performance levels are uptime, availability, reliability, and responsiveness. The agreement may also contain information relating to actions if the system goes down or an interruption occurs. The SLA protects the vendor and provides reasonable expectations for the client.

Often the SLA outlines provided services and responsibilities that fall to the client and vendor. It often includes a guarantee and warranty with remedies for failures. In many cases, financial penalty clauses are incorporated into the agreement. For some vendors, these are added to the account as credits rather than an actual payout.

In many organizations, the SLA has moved beyond the scope of the IT specialist and involves the oversight of a legal team. Often, organizations depend on cloud-resources for many mission critical items including protecting sensitive company data. Therefore, intellectual property protection, data security and other important concerns become part of the SLA.

While a legal team in an organization may influence policies and statements related to protecting intellectual property and organizational data, IT specialists may focus more on the levels of service and ensuring these meet enterprise requirements for response time, customer transaction processing, and other elements. In the end, the SLA should spell out clearly defined deliverables relevant to the IT function of an organization.

SLA Sources

SLAs come from different sources. Most are standard and provided by a cloud vendor. This is particularly the case when dealing with Amazon, Google, or Microsoft. A small business may not have much leverage when it comes to SLA definitions and development. That changes for big organizations where SLAs may be jointly developed according to best practices and changed to satisfy client needs (Diez and Silva 2013). As such, the SLA may emerge through negotiation. Once both parties are satisfied, the final agreement should be reviewed and approved by legal counsel (many organizations involve a legal representative throughout the process).

Other SLAs may be specified initially by the company seeking a cloud vendor. This could be part of the RFQ (Request for Quotation) package sent to various providers or opened for bidding. In this case, the purchaser may specify downtime requirements or service needs, and the vendor will supply a bid that either agrees with the terms or provides alternatives the purchasers can either agree to accept or refuse.

What Should SLAs Cover?

An excellent article by Thomas Trappler describes important aspects of SLAs and what they should cover at a minimum. He summarizes by suggesting a contract (Trappler 2010):

- Codifies the specific parameters and minimum levels required for each element of the service, as well as remedies for failure to meet those requirements.
- Affirms institution's ownership of its data stored on the service provider's system and specifies rights to get it back.
- Details the system infrastructure and security standards to be maintained by the service provider, along with rights to audit their compliance.
- Specifies rights and cost to continue and discontinue using the service.

SLA Components

An SLA generally comprises two areas: management features and service levels. Management features focus on several areas. First, standards, measurement techniques, methods for collecting data, report types, and report frequencies should be spelled out. These items provide the information used to define service levels. Other items on the management side relate to conflict resolution, SLA update processes, and finally an indemnification clause. This clause protects the client (as opposed to the cloud provider) from litigation brought about by service failures. This is not the remedy used to assess penalties for performance failure but rather for outside, third-party lawsuit protection against those who are not direct parties to the agreement. For instance, imagine that a data breach exposed 10,000 customers' credit card numbers and each of them suffered financial loss. The indemnity clause would protect the cloud client from those damages since the fault lay with the vendor.

The second part of an SLA deals with service elements which we have already mentioned. These specify the exact services provided, what is not specifically provided, conditions for service, time frames for service, responsibilities of each party related to service items, penalties for failures, metrics for service levels, and cost/service tradeoffs.

More on Indemnity

An indemnity clause is very important to the cloud service client and protects against lawsuits caused by cloud vendor actions. Most standard cloud provider SLAs do not include an indemnity clause but an organization about to enter a cloud service contract should demand that one be included. Many lawyers have experience with indemnity clause wording and can quickly put one together. The cloud vendor may seek to negotiate on this item, but it is very important and potentially very expensive considering the scope and number of recent data breaches.

SLA Metrics

Since an SLA defines a tangible service, most will spell out specific, measurable benchmarks to define what constitutes high quality service outcomes. The terms are defined prior to signing an agreement to remove any doubt regarding expectations of both parties. For instance, service levels may be measured using:

- Mean time between failures (MTBF).
- Mean time to repair (MTTR).
- Downtime per year or other time frame.

Most metrics should ensure that poor performance by either vendor or customer is not rewarded. The vendor may not be able to meet objectives if the client does not provide certain information in a timely fashion. And, of course, the vendor may not meet expectations if the service does not operate as claimed.

One issue regarding SLA compliance metrics relates to the source. In most instances, the cloud vendor provides the metrics as part of their platform. Often these are in the form of performance statistics and are visible through an online dashboard or in automatically generated reports. Most vendors rely on the cloud services customer to analyze and determine whether the contractual level of service has been met. Although, in some cases this is automated and lower levels of performance may result in service credits automatically being added to the account.

It is possible for the customer to utilize a third-party piece of software to monitor system metrics as well. If the customer desires to have an outside service used for metric determination, this may have to be established prior to the SLA being signed. Vendors may wish to have control over this aspect of the contract and may not agree to change to a third party if they typically use their internal systems for measurement.

In either case, it is important that cloud customer and vendor agree to the method for generating performance metrics during SLA negotiation to avoid problems later. And, even though most cloud vendors prefer to use their own metrics, if a customer has a mission critical computing need with goals that must be met, many experts recommend procuring a third party, either software or consulting company, to perform the metric collection and analysis. This removes misunderstandings that may arise on either side and provides an objective set of performance measures.

Metrics used to determine whether the terms of service are met can take many forms and often depend on cloud services used by the customer. Many items can be monitored from a cloud providers' perspective, but only those relevant to a customer's needs should be included as part of the SLA. Most vendors have a recommended, best practices set of metrics, but these will probably be favorable to them. A cloud services customer should understand their system needs and ensure the metrics are consistent with meeting business goals for the platform.

For that reason, it is best to keep the metrics at a minimum while still ensuring services meet needs. This helps avoid confusion when determining whether service levels meet contractual obligations and prevents the vendor from spending unnecessary time and resources meeting a meaningless benchmark.

Whenever possible, use automated monitoring and data collection and compile results into a report. Reports can be periodic or generated when service levels are not met. Many IT administrators find that data analysis, when not automated, takes a great deal of time and often is put on the backburner and forgotten.

As mentioned earlier in this section, several different metrics can determine if the cloud system is working well. Among these are service availability metrics, service quality metrics, security metrics, or key business performance metrics. We will look at quick examples of each. But, keep in mind that metrics should support the organization's goals for system use and may require custom development to ensure the system works as required.

Service availability metrics: This is a common metric used to assess vendor obligations. Basically, the vendor guarantees the service uptime. Generally, higher availability costs more because additional redundancies must be built into the system. In these cases, downtime becomes the time required for a redundant system to take over operations should a failure occur. Several cautions exist here. First, a 99.5% uptime (if that is the guaranteed level) sounds very good but must be considered within a time frame. For instance, imagine that the system works well all year apart from one failure, and this failure happened during Cyber Monday when the cloud services customer hoped to sell a huge amount of their product. Here is a simple calculation:

1 yr of time = 24 hrs × 60 minutes per hour × 60 seconds per minute × 365 days.
So, 1 year is about 31,536,000 seconds.
If we multiply that by the 99.5% uptime figure we get: 31,378,320 seconds.

If we subtract the uptime from the total time we get: $31,536,000 - 31,378,320 = 157,680$ seconds downtime allowed per year under contract. If spread over the year, that is only 0.5% of the overall possible usage time. But, imagine if the contract only specified uptime and did not specify any time period. That means the entire 0.5% downtime could occur at once and no penalty would be assessed. So, if that happened on Cyber Monday, it could mean that:

157,680 seconds / 60 seconds per min / 60 minutes per hr = 43.8 hours of downtime

If that occurred, the contract agreement would be fulfilled by the vendor. For a company that depends on uptime during Cyber Monday, that could be a disaster, because if the system were down for 43.8 straight hours, Cyber Monday (and most of Tuesday!) would be gone!

Imagine that your organization provided data feeds for the NFL Super Bowl broadcast and it went down during that critical time frame. Worst-case scenarios are easy to generate. In our prior case, if uptime was guaranteed to be 99.9%, an outage of 8.76 hours could still occur. If the rate was at 99.99% then the length of outage could be 52.56 minutes. So, the dilemma becomes apparent. In these cases when mission critical times are at hand, the vendor needs to supply 100% uptime or at least guarantee a maximum outage time per incident.

So, availability time, while a useful measure may need to be tied into a specific time slot with maximum continuous outage times specified. In fact, many e-commerce vendors or those that deal with highly time sensitive services require very high availability levels and very low maximum outage times. It is not unusual for a vendor such as these to require a 99.9999% availability in their SLA and demand additional performance guarantees during key times. Other metrics include:

Service performance metrics: Service performance speed is another area often considered in the development of metrics. This is a tricky metric because multiple factors are involved. For instance, if web services or other activities that require an internet interface are included, part of the response could be outside the vendor's control. Performance metrics may look at items like how long database reads or writes take; time required to provision a new VM; number of retries a database attempt gets; and, so forth. Often, these are custom metrics closely aligned with business goals of the cloud customer.

Service quality metrics: Some cloud customers are less worried about availability and are more worried about accuracy of storage, backups, web services, and other items. Imagine your organization performs scientific tests for cancer patients and needs to store results in a database ultimately used to determine medical treatment. In this situation, defective storage or failed backups could cost someone's life. So, a metric that represents this as an important activity would be more meaningful. Another metric could be related to coding errors or mistakes in services provided. If the cloud vendor develops scripts for automation or orchestration, a good requirement is that these be error-free. Production planners and quality management operations in organizations can provide advice about these sorts of metrics.

Security metrics: For other organizations, privacy and security-related concerns are paramount. For example, a credit card company may include a security-related metric in its SLA. These would have to be custom developed to match the use case but could include data breaches, virus infections, and speed at which software security updates are completed. As mentioned earlier in this chapter, indemnity clauses are recommended for organizations with high security needs and should be tied into these metrics.

Key business performance metrics: The best metric may be a key business performance related item. This would help cloud vendor and customer work together to achieve a desired outcome and enhance a partnership. The problem becomes one of where responsibility starts and ends. It may be difficult to determine what part of a key business performance index is provided by a cloud vendor. But, if this can be determined during the SLA negotiation, the result could be good for both parties.

Overall, metrics are in place to ensure systems operate as expected. The goal, for both vendor and customer, is an equitable metric that encourages business-enhancing outcomes and avoids any unnecessary costs and wasteful resource usage. Another consideration is to ensure selected metrics are within the cloud vendor's control. Do not measure items that cannot be changed or may be related to current state-of-the-art in technology. The SLA agreement must be realistic and ensure a working partnership emerges. Once metrics are selected, develop a realistic benchmark. It may be necessary to run systems and test important elements before finalizing these values. If historic data is not available, adjustment clauses may be needed at a predefined time.

Other Performance Considerations

The SLA may also list exclusions and exemptions. This simply means that at times, the vendor will not have to meet the acceptable performance levels. Since this means service will be diminished, the cloud provider should offer a specific list of why, when, and for how long these exclusions last. Generally, this includes scheduled maintenance time and may be negotiable to ensure limited disruption to the customer. The clients should carefully review exceptions and exclusion clauses and be willing to question these to ensure understanding. Times for maintenance can be adjusted to coincide with business practices.

Another consideration for performance testing involves the timeframe that a cloud customer plans to measure the cloud provider's compliance. Some agreements allow for a specific monitoring period, others a periodic time, and yet others permit continual monitoring of key metrics. This should be clear in a contract.

Performance Failure Penalties

Having a sound metric and a working relationship with a cloud vendor is essential. However, even in the best of circumstances, things can go wrong. Be sure to define penalties and resulting actions before a problem emerges. As mentioned before, a commonly used penalty for system shortcomings or failures could be service account credit. Even though this is common, setting an equitable rate for crediting the account might be necessary during negotiation. Other forms of penalties also are possible. Some cloud customers prefer direct compensation. The amounts could be tied to revenue loss during service outages. This incentivizes cloud providers to ensure uptime during peak customer times. Other penalties range from outside oversight to covering costs to switch to a new vendor.

Another clause found in some SLAs is called *earn-backs*. Earn-backs are the right to reduce service credits if actions are accomplished by the cloud vendor. These may not cover

---

9 Ownership of data

At all times, the Client will remain the owner of the data that is stored and managed in the Adlib databases. The Client will have full custody over the content of their Adlib database and Client specific digital assets such as images and electronic documents.

Figure 6.4 Sample data ownership clause from a publicly available SLA from Axiell ALM Cloud Service in the Netherlands.

an entire penalty amount but do incentivize cloud vendors to fix issues in a timely fashion. Many cloud customers that include earn-backs limit these to a narrow timeframe after penalties have been assessed to assure quick resolution of issues.

SLA Data Ownership Clause

Data storage is a primary SLA focus for many organizations. This creates important issues from a contractual standpoint, particularly if the cloud customer has sensitive or mission critical data. The following become very important:

- Who owns the data?
- Where is the data physically located?
- What happens to the data upon contract termination?
- What happens should a data breach occur?
- How are government or other requests for data access managed?

Data Ownership

Although this does not seem like it should be an issue, it is from a legal perspective. In many cases all of an organization's data will reside on a cloud platform which means it physically resides on another organization's hardware. The SLA, therefore, needs to clearly state that the cloud customer owns the data. Of course, it is possible to encrypt the data prior to cloud storage using techniques we mentioned earlier so the cloud provider cannot access the data's content, but ownership of the data needs to be clearly and unequivocally defined in the contract.

Most SLAs use standard language for this requirement. Figure 6.4 provides an example from the Axiell ALM Cloud Service – Service Level Agreement. As shown, the cloud vendor ensures the data stored on their cloud remains the property of the customer. It is also important to consider ownership of data created within the cloud if visualizations or other types of processing takes place.

Data Location

Many legal issues have emerged due to data location. The laws of countries and jurisdictions vary, and legal problems concerning data migration, import, export, and storage can result if a data center or backup is physically located in a country other than your own. For these and other reasons, SLAs should state the physical location for data storage. This can become very complex because a cloud provider may hold backup, production, and replication copies in different locations. It is even possible that data will be split among data centers located in different countries. Some nations have different forms of copyright laws and view data ownership differently. For instance, in the US, data typically belongs to the entity that collects it, whereas, in Europe, in many cases, data belongs to the individual whom the data describes. The US and China, for example, have very different copyright protection laws. And, in some countries, it is illegal to export data without special permission. Other questions arise as well. For instance, export controls exist regarding data transportation to certain nations. Without careful oversight and help from the cloud vendor, laws could be violated inadvertently.

Countries with Strong Data Privacy Laws

Organizations or individuals may wish to store data in locations where government access is limited. This could be for nefarious reasons (e.g. drug cartels or Mafia-based systems) or just because an organization feels threatened (e.g. political parties or religious groups). The following three countries offer good data protection and are good hosts for VPN services.

Switzerland: Of course, Switzerland has long been the paragon of neutrality. It maintains a longstanding policy of not being party to political agreements, extraditions, or other forms of international interaction. Switzerland also provides a good location for data storage and is a strong advocate for data privacy. It has a Federal Act on Data Protection law and generally refuses to allow access to its citizens' data. It is a pro-consumer country. It does, however, have an agreement with the US government that allows data access through a strict subpoena process related to banking fraud and terrorism.

Malaysia: This country has very strong data privacy laws which protect data owners. It also has a law called the Malaysian Personal Data Protection Act to keep its citizens' data private when possessed by others. It has heavy fines for violations. This is a good place for data storage if the data set does not contain the personal data of Malaysian citizens.

Iceland: This geographically isolated nation has no agreements for data sharing or access by the US or other countries. For US-based companies, this is a great data storage choice located between Europe and North America. Iceland has a long history of challenging foreign government data access requests and takes its custodial role very seriously.

Of course, no data is 100% guaranteed to never be accessed by government agencies, particularly when crime or terrorism is motivating the access.

When it comes to US government data storage, cloud providers must ensure data is stored within the continental US. For instance, Azure's cloud includes the following guarantee, "Azure Government offers data centres with round-the-clock monitoring, operated by screened US citizens. All data, app and hardware residency remain in the continental United States." Similar features are offered by Google and Amazon for their government cloud services.

Data Disposition

Data disposition clauses deal with vendor actions after a contract is terminated. It is important to define this in an SLA for several reasons. First, an organization does not want to be indefinitely tied to the same vendor. There needs to be an orderly way out. Second, there must be an economically feasible and realistic way to transfer to another vendor. For example, if a cloud provider only permitted a client to remove files one-by-one using FTP, it could become impossible to migrate to another platform.

Migration may be necessary for several reasons. While some relate to vendor performance or costs, often an organization will need to migrate because they have consolidated or been purchased, and systems will be integrated.

Several issues to consider include the media that migration will use, the time delays involved, and who completes the migration. In some situations where emergencies could occur or time-sensitive data exists, it might be necessary to develop acceptable procedures and add those to the SLA. An example of contract language provided by Internet2 follows:

> Upon request by Customer made before or within sixty (60) days after the effective date of termination, [Vendor] will make available to Customer for a complete and secure (i.e. encrypted and appropriated authenticated) download file of Customer Data in XML format including all schema and transformation definitions and/or delimited text files with documented, detailed schema definitions along with attachments in their native format. [Vendor] will be available throughout this period to answer questions about data schema, transformations, and other elements required to fully understand and utilize Customer's data file. After such sixty (60) day period, [Vendor] and its hosted service provider shall have no obligation to maintain or provide any Customer Data and shall thereafter, unless legally prohibited, delete in such a manner as prevents recovery through normal/laboratory means, all Customer Data in its systems or otherwise in its possession or under its control.[1]

This sample clause provides guidance to the vendor and client that spells out terms of data return following contract termination. Notice it also talks about customer data format and describes that following return of the data, and after a 60-day waiting period, it should be deleted. Other related contract details should specify that a cloud vendor must destroy all copies of the data without sharing that data with outside parties in any form. Sometimes cloud customers also include language that permits them to hire an auditing service to ensure data has been deleted.

Data Breaches

An area receiving a great deal of press relates to data breaches or situations where hackers illegally obtain data from a cloud provider or other source (Kolevski and Michael 2015). As mentioned earlier, many contracts include an indemnity clause that shifts the cost of a data breach to the cloud provider should it be their fault. This indemnity clause ensures any lawsuits are covered by the vendor rather than the cloud client. Even with an indemnity

---

1 From Educause, request by Customer made before or within sixty. © EDUCAUSE.

clause, it is important to consider the actions needed should a worst-case scenario occur. In other words, what is expected of the vendor and what actions will be managed by the cloud customer. The actions to be taken depend, at least in part, on the nature of the data stored. For instance, if credit card data, health data, or other personally identifiable data exists, the situation becomes more dire.

One important fact outlined in most contracts relates to disclosure. A vendor generally is required to disclose when a data breach occurs within a specified timeframe. They also must indicate whether a verified breach occurred or if a breach may have occurred. Often, a report of circumstances leading to the breach is required or, in some instances, a face-to-face meeting is required. Cloud customers may wish to include language that prohibits mentioning them in a press release or other public announcement, preferring instead to make those sorts of public statements themselves. This may not be possible due to laws governing the release of details concerning data breaches, particularly when personally, identifiable data exists. Sample indemnity language as provided by Internet2's website follows:

> [Vendor] shall defend and hold Institution harmless from all claims, liabilities, damages, or judgments involving a third party, including Institution's costs and attorney fees, which arise as a result of [Vendor]'s failure to meet any of its obligations under this contract.

And:

> [Vendor] shall indemnify, defend and hold Institution harmless from all lawsuits, claims, liabilities, damages, settlements, or judgments, including Institution's costs and attorney fees, which arise as a result of [Vendor]'s negligent acts or omissions or willful misconduct.[2]

Governmental Access Requests

Another issue that may come up in an SLA involves the cloud provider's actions in circumstances where governments, law enforcement, or other regulatory authorities need access to data. Most SLAs contain governmental regulation compliance languages. For example, from Internet2:

> [Vendor] and/or its agents or employees agree to comply with all laws, statutes, regulations, rulings, or enactments of any governmental authority. [Vendor] shall obtain (at its own expense) from third parties, including state and local governments, all licenses and permissions necessary for the performance of the work.[3]

---

2 From Educause, shall defend and hold institution harmless © EDUCAUSE.

3 *This clause is licensed under a Creative Commons Attribution-NonCommercial-ShareAlike 4.0 International License(CC BY-NC-SA 4.0).* It was retrieved from: https://spaces.at.internet2.edu/display/2014infosecurityguide/References+to+Third+Party+Compliance+With+Applicable+Federal%2C+State%2C+and+Local+Laws+and+Regulatory+Requirements

---

Internet2 Website

The Internet2 is a website, supporting Higher Education administration. It catalogs open access resources such as SLA statements, data privacy information and other items that can help an IT administrator ensure all aspects of an agreement are considered. Many of this book's sample SLA statement are found on their website. A link to an index of their resources is: https://spaces.at.internet2.edu/display/2014infosecurityguide/Toolkits.

As shown, both client and vendor expect to follow all laws within the terms of service. Occasionally, this requires a vendor to turn over data for a criminal investigation of an individual whose credit card or other data resides on their system. There also could be circumstances where another client of the cloud provider is under investigation and as a result a government agency has access to all portions of the provider's cloud. In all related situations, the cloud customer should ensure the provider is responsible for contacting them as quickly as possible. Ideally, that would be prior to releasing data but some court orders could prohibit that from occurring.

SLA Revisions

SLAs are not immutable. Businesses grow and change. New services are added. New technologies emerge. New laws come into force. So, an organization must be prepared for its SLA to be periodically updated. No foolproof way to anticipate the future exists but an organization prepares by having a solid SLA and then working closely with their cloud provider to periodically update the agreement to benefit both parties.

Many times, changes will be positive, for instance, increasing access speed or updating hardware or software systems. Some vendors will provide incentives for customers to move to new software versions. Other times, the technical environment will mandate it for security or other reasons. It is important to hold periodic reviews and determine what changes are needed and then work with the vendor to move forward.

Transferring SLAs

If an organization is purchased or merged in an acquisition, the SLA may have to be renegotiated or reworked. A cloud vendor is under no obligation, unless specifically noted in the SLA, to keep the same agreement under new circumstances. Most cloud providers are happy to work with the changed organization to ensure it remains a customer and therefore may continue to honor the SLA or provide a reasonable update to it. The cloud customer should get the new agreement in writing to ensure the SLA remains in force and is not subject to any unexpected changes in status.

More on SLAs

Most of the discussion on SLAs in the prior sections has focused on private cloud vendors that offer special services to larger institutions or businesses that will provide substantial revenue. For small businesses or individual cloud users, the SLAs usually are standard agreements that most public cloud providers offer. This is particularly true for cloud storage service providers. Since they serve large numbers of smaller clients, they offer generic service sets to everyone. The following general guidelines apply to evaluating standard SLAs:

- Read the agreement carefully prior to subscribing. The SLA may impact choice of a cloud vendor.
- Have an IT specialist from within the organization, or technical consultant from an uninvolved organization, review the SLA for any details that might not be clear. Again, finer points might influence vendor selection.
- Ensure the legal framework for the SLA is sound and fair. Get legal counsel to review the agreement and provide advice.
- Ensure you have contingency plans in case the vendor goes out of business or has other difficulties. This may mean having an onsite backup periodically made or having another solution.

The SLA is a contract and it sets the expectations for both client and vendor. It protects the cloud service provider from unfair demands and ensures the client is treated in a predictable, fair manner. The document forms the basis for a partnership and when issues do arise, it provides a basis for resolution.

Billing

A side benefit of cloud computing is its approach to billing. Instead of paying large, up-front costs, most cloud users start the service and pay a periodic bill much as they would with any utility. Nearly all cloud service providers use the *utility billing approach* to charge for their services. This provides a fair way for vendors to bill for elastic operations. The more a customer uses, the more they pay. This, as mentioned in Chapter 1, replaces the old model for capital investment in large computing infrastructure systems.

Just to clarify, here is how utility billing works. Think about your home for a moment, nearly everyone has electricity that powers many aspects of their lives. In most rooms, you can find an outlet and plug in whatever you need to operate. When you make your morning coffee, the coffee maker uses a small amount of power that is recorded in your electric meter. If you make toast, the same thing happens with your toaster. Unless you use the electricity, you do not pay for it. Cloud billing works the same way: unless you use the cloud for a purpose, you do not pay. So, if you run a database query, you pay a small amount for that. If you create a virtual desktop, that is metered for another charge and so forth. The cloud is like a large utility that only costs when you use it.

So, the cloud can be likened to the electric power grid. The consumer is not too worried about where the power comes from as long as it works. Energy may come from the Edison Sault Power Hydroelectric plant, from coal-based resources, or any number of other sources: wind, solar, or nuclear. Electric companies buy and sell excess electricity depending on customer needs. The same could be true in the cloud. Excess computing power can be sold to partner firms and the need for extra power may be acquired from partners. Companies in this business attempt to optimize their profit by buying/selling according to various schemes which has the net effect of lowering everyone's prices (at least that is the idea!).

Although like electric utility billing, differences exist regarding cloud computing billing – the primary being complexity. The average electric company customer sees a single monthly bill that provides a straightforward number representing overall usage. Although this would be possible for cloud providers, their offerings are far more complex, and their customers need to understand computing resource consumption at a very granular level. This means the accounting staff in most businesses need to tie exact costs of service to parts of the organization using those resources. This helps with budgeting, evaluating alternatives, and deciding which parts of a firm are profitable and which are not.

Think of this in terms of your home. Instead of a single monthly usage number, you want to know how much your toaster operations cost, how much electricity watching television consumes, or how much it costs to power your coffee habit. This granular level of understanding more closely matches what cloud clients need to know. And, unlike your electric bill which comes at the end of the month, many organizations want to know what they spend on computing resources in real time. This avoids shock and helps manage budgets in ways that keep costs under control.

Fraunhofer Institute for Microelectronic Circuits and Systems

Electric utility billing soon will be more granular and more closely match cloud computing billing approaches. Researchers at Fraunhofer Institute for Microelectronic Circuits and Systems IMS in Duisburg, Germany, recognized that electric devices have identifiable signatures that enable their software to determine the portion of electricity each uses. This information feeds into their reporting system that breaks down usage by item. So, in the future, you will know how much electricity you used watching television, charging your phone, making toast, and, yes, even catering to your coffee habit! The desired effect will encourage manufacturers to reduce device-specific electric consumption and encourage customers to replace devices that are heavy consumers of power. In the end, an overall cost of usage will become apparent and influence what you buy for your home.

Amazon Billing

Amazon is a leader in developing billing approaches for cloud computing and an advocate for utility-style billing. As such, Amazon charges for what a client uses but, as might be expected, many variables and services go into the mix resulting in a complex monthly bill. AWS provides a console/dashboard to help decipher the result. Among its features are:

1. *Cost Explorer* to enable a client to track and analyze AWS usage. This is free for all accounts and offers some customization.
2. *Budgets* which is a feature to manage budgets and set cost alerts for overruns and so forth.
3. *Bills* to see details about current charges.
4. *Payment History* to view information about past payments and transactions.

In addition, Amazon provides a wide variety of billing reports to help clients understand their usage. The overall billing system is well developed, but it is complex. In fact, a 300+ page guide provides information about using the billing system (Amazon Web Services, 2020).

Third Party Billing Tools

Several organizations provide billing tools to help with cost management in cloud computing. We mentioned several in the "Cost Monitoring" section of this chapter. Among other tools that specialize in this area are:

ScienceLogic Cloud Management Tools: These tools provide a range of cost management and billing support tools.

Cloudability: These tools focus on cloud cost and usage data for organizational decision-makers. It provides specialized reports focused on operations, accounting needs and management teams.

Summary

Without a doubt, cloud computing has a complex business model that matches well with a modern businesses' information requirements. Managers and accountants must be able to understand IT costs from a usage perspective and tie that into requirements for elasticity, resource demand changes, flexibility, agility, and smart planning. The cloud moves businesses from a capital investment model for computing infrastructure to one that more closely matches a utility company. As resources are consumed, costs are incurred. The cloud is largely a pay-as-you-go model.

This chapter has reviewed several approaches, tools, and practices that support business operations in areas related to the cloud. We examined monitoring and console tools that provide IT specialists and managers with ways to understand and predict usage. We saw how usage directly ties to organizational entities, and how monitoring enables IT specialists to ensure resources operate as desired. Issues with capacity and other important attributes can be "watched" either directly or through automated management tools that send out alerts. Monitoring also aids in the security process and ensures the correct resources are accessed only by authorized users. We saw how tools help minimize waste and prevent zombie resource instances from driving up costs.

This chapter also covered SLAs. These are binding contacts between cloud vendors and their clients. An SLA usually describes conditions for use and minimum level of service and performance the cloud vendor provides. Among service and performance level requirements are uptime, availability, reliability, and responsiveness. SLAs generally contain information describing actions should the system go down or an interruption occurs. The SLA protects the vendor and provides reasonable expectations for the client.

Finally, we looked briefly at how cloud providers approach billing. We saw that nearly all cloud service providers use the utility billing approach to charge for services. This is a fair way for vendors to bill for elastic operations. The more a customer uses, the more they pay. Cloud clients tie this granular information directly to operations to assess and track costs.

References

Amazon Web Services (2020). *AWS Billing and Cost Management: User guide*. United States: Seattle.

Diez, O. and Silva, A. (2013). Govcloud: using cloud computing in public organizations. *IEEE Technology and Society Magazine* 32 (1): 66–72.

Kolevski, D. and Michael, K. (2015, October). Cloud computing data breaches: A socio-technical review of literature. In: *2015 International Conference on Green Computing and Internet of Things (ICGCIoT)*, 1486–1495. IEEE.

Trappler, T.J. (2010). If it's in the cloud, get it on paper: Cloud computing contract issues. *Educause Quarterly* 33 (2).

Further Reading

Barona, R. and Anita, E.M. (2017, April). A survey on data breach challenges in cloud computing security: Issues and threats. In: *2017 International Conference on Circuit, Power and Computing Technologies (ICCPCT)*, 1–8. IEEE.

CIO Council and Chief Acquisition Officers Council (2012). Creating effective cloud computing contracts for the Federal Government: Best practices for acquiring IT as a service. https://s3 .amazonaws.com/sitesusa/wp-content/uploads/sites/1151/2016/10/cloudbestpractices.pdf.

Compagnucci, M.C. (2020). *Big Data, Databases and "Ownership" Rights in the Cloud*. Springer.

Dhirani, L.L., Newe, T., and Nizamani, S. (2019). Federated hybrid clouds service level agreements and legal issues. In: *Third International Congress on Information and Communication Technology*, 471–486. Singapore: Springer.

Emeakaroha, V.C., Brandic, I., Maurer, M., and Dustdar, S. (2010, June). Low level metrics to high level SLAs-LoM2HiS framework: Bridging the gap between monitored metrics and SLA parameters in cloud environments. In: *2010 International Conference on High Performance Computing & Simulation*, 48–54. IEEE.

Hurwitz, J.S. and Kirsch, D. (2020). *Cloud Computing for Dummies*. Wiley.

Jones, S., Irani, Z., Sivarajah, U., and Love, P.E. (2019). Risks and rewards of cloud computing in the UK public sector: A reflection on three Organisational case studies. *Information Systems Frontiers*: 1–24.

Stephen, A., Benedict, S., and Kumar, R.A. (2019). Monitoring IaaS using various cloud monitors. *Cluster Computing* 22 (5): 12459–12471.

Tarawneh, A. and Abdulrahman, A.A. (2019). Analysis and management of risk related to using cloud computing in business: employee's perception. *International Journal of Management* 10 (6): 20–32.

Terfas, H. (2019). The analysis of cloud computing service level agreement (SLA) to support cloud service consumers with the SLA creation process. Doctoral dissertation. École de Technologie Supérieure.

Walia, M.K., Halgamuge, M.N., Hettikankanamage, N.D., and Bellamy, C. (2019). Cloud computing security issues of sensitive data. In: *Handbook of Research on the IoT, Cloud Computing, and Wireless Network Optimization*, 60–84. IGI Global.

Zeng, X., Garg, S., Barika, M. et al. (2020). SLA management for Big Data analytical applications in clouds: A taxonomy study. *ACM Computing Surveys (CSUR)* 53 (3): 1–40.

Website Resources

Cost and Monitoring Software

Amazon CloudWatch: https://aws.amazon.com/cloudwatch
AppNeta: https://www.appneta.com
AppRiver: https://appriver.com
AWS Cost Explorer: https://aws.amazon.com/aws-cost-management/aws-cost-explorer
BMC TrueSight Pulse: https://www.bmc.com/it-solutions/truesight.html
CA Technologies: https://ca.com
Cloud Cruiser (HPE): https://www.hpe.com/us/en/services/cloud-cost-management.html
Cloudability: https://www.cloudability.com
CloudCheckr: https://cloudcheckr.com
Datadog: https://www.datadoghq.com
Dynatrace: https://www.dynatrace.com
Exoprise: https://www.exoprise.com
Google Cloud Operations: https://cloud.google.com/products/operations
LogicMonitor: https://www.logicmonitor.com
Microsoft Azure Monitor: https://azure.microsoft.com/en-us/services/monitor
Rackspace Monitoring: https://www.rackspace.com/openstack/public/monitoring
Retrace by Stackify: https://stackify.com/retrace
ScienceLogic Cloud Management Tools: https://sciencelogic.com
Spiceworks: https://www.spiceworks.com
Teevity: https://www.teevity.com

Zombie Instance Management Software

RightScale: http://www.rightscale.com
Total Cloud: www.totalcloud.io

7

How Are Business Applications in the Cloud Managed Safely?

A very important concern in the modern world relates to security and access control. Mission critical data and the need to connect systems across distances require that IT specialists be particularly vigilant regarding who can use, see, or change online resources. Security concerns change for a firm when it moves its IT services into the cloud. A common misconception is that suddenly, security becomes solely the concern of the cloud provider and the cloud client is relieved of that responsibility. While that is partially true, organizations need to take a proactive role in ensuring their resources are kept safe and secure. Security becomes a shared responsibility between cloud vendor and cloud client.

The process begins with recognition that cloud hosting has key differences from traditional, on-premise hosting. For example, information and services are housed on a combination of virtual and physical servers accessible via Internet or other network versus physical servers located onsite within an organization. The cloud makes some forms of security easier, for instance, distributed data across multiple, redundant servers protects against data loss and hardware failure. But the cloud also invites other forms of trouble, such as data thieves, malicious hackers, malware, ransomware, and other issues that deal with having data stored on someone else's remote, physical hardware.

The tools, procedures and concepts that cloud computing uses to ensure safety are the primary focus of this chapter. We start out with a general overview of cloud vulnerabilities then move into security architecture for each cloud model. Finally, we look at access and identity control.

Cloud Vulnerabilities

IT administrators often let out a sigh of relief when their operations move into the cloud. Many security headaches associated with on-premise solutions disappear. For instance, hardware worries, access to computing facilities and so forth are now managed elsewhere. This is not to say that all issues magically disappear. In fact, the cloud is vulnerable to problems and issues that on-premise solutions are not. The problems, however, are manageable. It is important to understand what security issues do exist and be ready to ensure appropriate measures are taken. Among unique cloud security issues are:

- IT administrators may lose the ability to manage user control and visibility because cloud vendors manage those services.

Cloud Technologies: An Overview of Cloud Computing Technologies for Managers, First Edition. Roger McHaney.
© 2021 John Wiley & Sons Ltd. Published 2021 by John Wiley & Sons Ltd.
Companion website: www.wiley.com/go/mchaney/cloudtechnologies

- Managing resources and services becomes more challenging due to virtualization and provisioning with resources that are not fully removed when not in use. This can result in unauthorized cloud services, malware infections or data leakage. This can be shortened to say: IT administrators may lose the ability to fully manage cloud resources.
- The widespread use of APIs to access cloud services opens new points of vulnerability and potential weak spots for malicious attacks and malware.
- Use of cloud resources by multiple tenants leaves open the possibility that overlap occurs in a shared public cloud environment.
- Data and other resources might remain in unexpected locations because these are spread among multiple storage devices in decentralized ways.

Besides the unique threats, cloud computing is still subject to problems that plague on-premise systems. These may include phishing, malicious insider actions, ransomware, malware, unsecured accounts, and unsecured data.

Cloud Security Architecture

Cloud security starts with ensuring sound architecture has been put into place for each level of infrastructure (Hashizume et al. 2013). We will look at IaaS, PaaS, and SaaS to determine the actions recommended for each cloud implementation service. In all cases, security falls into a shared responsibility model which means the cloud provider and cloud client need to work together to keep resources safe and secure. We covered SLAs in the prior chapter and this legal contract generally spells out the cloud provider's role in security. The customer needs to be aware of where the vendor's responsibility ends and theirs begins. Most cloud vendors approach this arrangement, the shared responsibility model, in this way:

- The cloud provider assumes responsibility for security of cloud resources including the infrastructure, storage, databases, networks, containers, VMs, and so forth.
- The cloud client assumes responsibility for use of resources in the cloud including software applications, data, services, operating systems, firewalls, user management, and so forth.
- Many cloud providers offer extra services to help manage aspects of the cloud client's responsibilities.

IaaS Security Architecture

IaaS is concerned with infrastructure and this creates a scenario where the cloud client assumes more security-related responsibility. Of course, if an organization is implementing its own private cloud, the IT staff assumes all responsibility for security. We will focus mainly on implementations that are using a public cloud service supplied by a vendor like Amazon, Microsoft, Google, or IBM. In most IaaS environments, the cloud customer has access to the underlying IT infrastructure. The customer uses templates and scripts to provision resources on demand and manage deployments and related security here. The cloud provider owns the hardware and data centers and manages those areas regarding security. Figure 7.1 provides a view of the shared responsibility.

Figure 7.1 Cloud security in IaaS: Shared responsibility model.

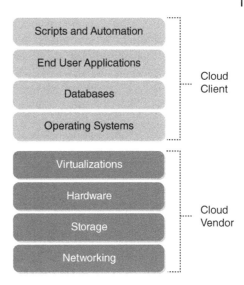

IaaS comes with three major security risks like those in on-premise systems but which are unique due to the operating environment. These risks include resource misconfiguration, vulnerabilities, and zombie systems. Appropriate pre-emptive actions can minimize the impact of these risks.

IaaS Resource Misconfiguration

Resource misconfiguration relates to how and where various cloud systems are deployed, and common mistakes that occur when this is done. In other words, when scripts are created for automation or orchestration, certain actions make a cloud resource more vulnerable to error, attacks, or other problems. An example includes putting file structures into default locations. For instance, when a web server is set up, a default folder structure is put into place. Those wishing to cause harm will be given an advantage if the default structure is used by the IT administrators. This does not mean a problem will occur, it is just more likely. In addition to using default folder structures, other administrators may use very basic or "guessable" structures. Default resource passwords and userIDs might be left on the system which provides an opening to anyone understanding possible intrusion pathways.

Another area for concern regarding IaaS is that it provides storage and networking resources for cloud networking. Often, these are accessible through APIs to enable end users to interact with cloud data and programmed services. APIs, by their very nature, are open to the outside world. This means that they provide a target to harmful hackers or others seeking a door into the system. Often APIs rely on custom developed software which may be subject to issues related to programming errors or known vulnerabilities.

IaaS Resource Vulnerabilities

Another consideration with IaaS relates to security issues. It is possible that incorrectly configured access permissions give unauthorized users access to sensitive, cloud-based information. Some folders might contain additional files with sensitive information useful for additional system malfeasance.

Another IaaS problem relates to scripts developed for automation and orchestration. It is possible that secure information is hardcoded into this computer code. This may include paths, userIDs, passwords, and other secure items in human readable form. Giving the wrong person access results in a situation called *leakage*. This is the unintentional dispersion of sensitive information. For instance, perhaps a friend of yours at another organization sends you a script they created to spin up a VM. Perhaps that script has file location information or even a password coded in human readable form. Maybe a month later that script is emailed to another friend, who in turn posts it out on Reddit or a help forum. Eventually, the wrong person sees it and puts the overexposed information to use.

It is possible to employ outside services, system auditors and use other approaches in attempt to find these sorts of issues, but it is difficult. Many experts recommend making cloud traffic and usage more visible. One way of doing this is to use *network packet brokers (NPB)*.

NPBs are devices that offer monitoring capabilities and tools that oversee network traffic at cloud endpoints. NPBs interact with other monitoring and security systems, and feed data from the network stack. This provides extra visibility into IaaS security within a cloud-based infrastructure. Best practice uses NPBs to direct traffic and data to the appropriate performance tools and security applications. Further, the NPB can log data to allow further and more in-depth analyses.

In addition to the NPB devices, IaaS requires other additional security measures:

- Use of Virtual routers
- Virtual web application firewalls that protect servers against malware and viruses.
- Virtual network-based firewalls used to guard cloud endpoints.
- Intrusion Detection and Prevention Systems (hardware and software).
- Segmenting Networks with special consideration given to areas connected to sensitive or highly targeted resources.

Example: Configuration Entry Test from Outpost24

Outpost24 (https://outpost24.com) is a cloud services organization that offers its clients, among other products, IaaS security testing. One of their recommendations is to run a series of security checks to preemptively find vulnerabilities. For example, they suggest using a *configuration entry test* to find high risk web resource configurations. Output from their tool would show something like this:

CIS-AWS – 1.1 Avoid the use of the "root" account
  Service IAM – Risk Level High
Description: Using the "root" account entity is dangerous and should be avoided, if possible. Users should practice "least-privilege," a technique where specific user accounts are created and assigned the minimum privileges necessary to complete their work. Additional privileges can be added to their account as their scope grows, but no user should have the limitless power of the "root" entity.
Resolution: Create an IAM user and assign the basic role or privilege necessary to perform daily functions.

IaaS Zombie Vulnerabilities

An IaaS area of security concern relates to a topic we have mentioned before: zombie resource instances. Tools from a variety of vendors, including native tools from cloud providers, offer ways to find these instances and help remove them before a crisis emerges. We described some of these in the monitoring sections of the previous chapter. Looking at the same area from an IaaS security perspective, several types of zombies are concerning. Among these are:

Zombie servers and zombie workloads: Often, servers or workloads are started as tests and then not deprovisioned afterwards. Or, they are created during development efforts and not properly removed by programmers. These instances can be security risks because they provide entry points into resources that may not be watched carefully.

Zombie/orphan storage: Like zombie servers, zombie storage can be remnants from testing or development that have not been deprovisioned. Storage issues also relate to disks not attached to the system in a formal way. This is known as orphan storage. If critical data were placed onto an unmonitored location, leakage or loss can occur.

Workload problems: Hackers can use cloud resources to set up mail servers, or programs that attempt to launch attacks or determine passwords. These are essentially rogue processes that should not be running. Again, tools can detect and terminate these sorts of activities.

Temporary or dormant resources: If resources are taken offline temporarily and a security update occurs during that time, the item may be out of date and vulnerable to attack. Again, tools can watch for these issues and ensure that situation does not occur.

PaaS Security Architecture

PaaS, as you may recall, is platform as a service. This means the cloud customer relies on the cloud provider to supply the hardware, some software, and provisioning capabilities for the system. In PaaS services, the cloud host takes on more security measures than in IaaS security. In addition to the areas mentioned in the IaaS architecture section, PaaS vendors generally offer security for logging in, monitoring API gateways, and restricting IP addresses. Often, PaaS providers offer cloud security features that ensure visibility, compliance, data security, and threat protection. These solutions are known as a *cloud access security broker* or *CASB*. Figure 7.2 shows CASB's place in an overall cloud security diagram. Here is a closer look at the four main components of CASB:

Visibility: The ability to see resources is a key concept when it comes to cloud computing. An organization loses system visibility when it uses PaaS or other cloud infrastructure because the cloud provider manages aspects of resource usage and user access. So, the client organization's IT administrators may not know which users are accessing data, storage, devices, or systems. CASBs help by providing services to enhance visibility which allow administrators to see who is using what. This is particularly helpful when it comes to key, sensitive resources. You may be wondering: how is this different from monitoring resource usage mentioned in an earlier chapter? It is not that different. Having a CASB system means the vendor manages monitoring and allows your organization to access generated data. The CASB also prioritizes and acts on incidents that occur.

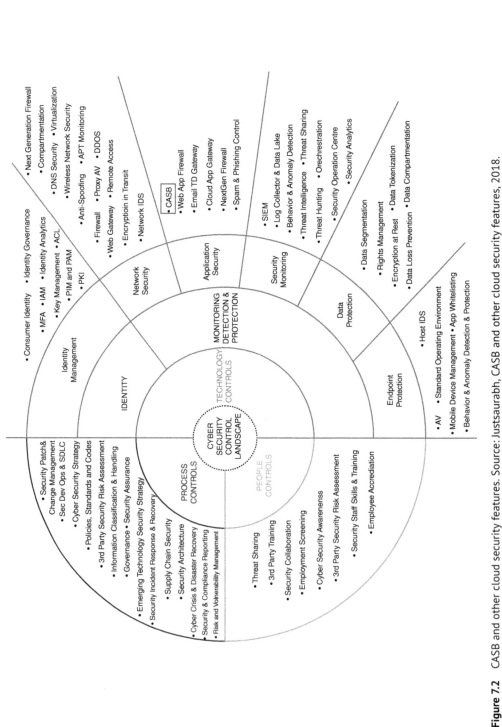

Figure 7.2 CASB and other cloud security features. Source: Justsaurabh, CASB and other cloud security features, 2018.

Compliance: CASBs address compliance issues. These could be laws, regulations, or organizational policies. Solutions provide compliance with regulations like HIPAA and FERPA in the USA and ensure system rules are updated as laws change. You may wonder how a CASB manages regulations. This is done in different ways. For a quick example, imagine an employee must take FERPA training before having access to sensitive, personally identifiable student data. The CASB service blocks access to any untrained users.

Data Security: CASB systems provide added protection for sensitive data. For instance, transmission of certain data may always require encryption. Other data must be accessed from specific IPs. Likewise, data may be accessible only with an administrators' specific permission and so forth. In general, CASB-enforced rules and protections ensure the correct people access critical data assets.

Threat Protection: Additional malware, firewall, and virus checking measures may be provided within the CASB framework. Particular attention may be given to critical resources and areas of the platform that contain sensitive data or systems.

CASB solutions primarily focus on application security (as shown previously in Figure 7.2). It integrates directly with cloud platforms and provides system use intelligence and enhanced visibility. It is elastic and useful in incident detection and resolution. CASB is not just a PaaS feature and is useful on most cloud platforms (e.g. SaaS, PaaS, IaaS). CASB forms a layer between the cloud client and cloud vendor but is not just an API-based tool. Instead, it is an integrated solution that takes actions at the platform level. Examples include revoking access or quarantining data.

SaaS Security Architecture

SaaS cloud providers take on a high degree of security-related issues for their customers. Often, these services are used by individuals or by small businesses with little or no internal IT functions at their location. Therefore, nearly all aspects of security must be managed by the cloud provider. Remember that SaaS clouds host software and data accessible via mobile devices or web browsers and this requires additional security measures.

As in PaaS systems, CASBs are used extensively by providers to monitor and discover problems or potential weaknesses related to security in SaaS systems. CASBs log activities, permit audits, monitor resources and uses, provide access control to data and services, and may enable an organization to manage its encryption. SaaS vendors also may provide IP restrictions and monitor security at API gateways. It is, however, up to cloud customers to determine who should have accounts using the software and systems; and to set users' access levels. Often, divisions of security roles are defined in SLAs. Figure 7.3 provides an overview of differing security responsibilities regarding SaaS, PaaS, and IaaS clouds.

Cloud computing is not less secure than traditional on-premise applications. In fact, many applications provide security solutions that far exceed traditional computing approaches. Among these are monitoring, standardization, ease of encryption, and CASBs. However, with these new advantages come challenges. IT administrators must truly understand the shared responsibility model applicable to their cloud computing infrastructure and work with vendors to ensure security at all levels. All cloud models, at a minimum should include the following features:

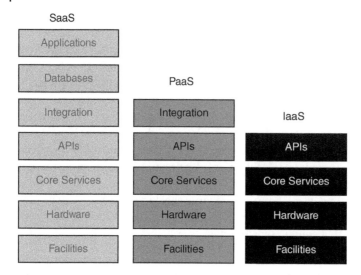

Figure 7.3 Typical vendor security responsibilities in different cloud computing models.

1. Use a single sign-in approach for accounts with services from multiple vendors to enhance logging, monitoring and visibility.
2. Use virtual firewalls and ensure these are updated regularly and vigorously.
3. Use virus checking, malware detection/prevention and data loss prevention tools.
4. Perform regular audits of security and usage patterns.
5. Add a security information and event management system (SEIM) and a denial-of-service protection package (DDoS).

Access and Identity Control in the Cloud

Determining who needs access to critical resources is an important aspect of cloud security for organizations. Not only do IT administrators need to know who requires logins, they must ensure logins offer access to resources appropriate to organizational positions. This can become a complex situation. In any organization, sets of people with different IT needs must access cloud resources for different reasons. Business policies and organizational configuration drive access decisions. Most businesses have several major groups requiring access. These include:

- *General Employees*: People that work for an organization and need general access to a set of common resources required to complete their jobs. This might include MS-Office, their health benefit software, email, calendars and so forth.
- *Managers*: These individuals will need access to organizational systems that relate to their jobs and need access to data about their directly reporting employees.
- *Business Administrators*: People at higher levels in an organization need access to strategic information, accounting reports, and systems that support their decision-making roles.
- *Customers/Clients*: People that use an organization's e-commerce system may need access to order entry systems and so forth.

- *Business Affiliates*: Business partners may need to see inventory levels or access information to support their roles.
- *Consultants*: Temporary and limited accounts may be assigned to people working with an organization.
- *Business Discipline Groups*: Salespeople might need access to software like Salesforce. Accountants need to access the accounting system. In general, people need access to specialized systems that support their specific job roles.
- *Groups for IT Automation*: In many cloud systems, accounts will be set up to help with automation and orchestration tasks. These are used by IT administrators to perform needed tasks.

In addition to the roles in the preceding list, other groups and special use cases are needed in any organization. So, policies must be in place to manage who can access what. This security practice is called *identity and access management* (IAM). In many organizations, the IT administrators work with other business leaders to develop a list of resources and a template for matching individuals to those resources. The person who oversees this role often is called a *business administrator* and is usually part of the IT staff.

The process can be simplified to some degree by creating groups that manage most user needs. This can be accomplished with a software tool like Microsoft's *Active Directory* or Google's *Cloud IAM*. In general, these utilities create an approved resource set for each business unit. A best practice is to place users into groups mirroring job responsibilities. Direct assignment of users to resources is poor practice because it makes access management a nightmare and may result in inappropriate allocation. Nearly every organization has special case individual users who do not fit any categories and must be managed separately. Resource templates are helpful because the same template can be reused for multiple groups or can be copied and modified as needed.

Identity Governance

As mentioned in the prior section, determining who should have access to which resources is a big challenge in most organizations. This task starts with IT administrators who work in conjunction with other leaders to develop an identity governance strategy. Many organizations have this in place prior to cloud migration but even so, policies need modifications and updates for cloud environments.

As a starting point, for all possible cases, accounts need to be listed in a central directory service. These services (e.g. Microsoft's Active Directory) facilitate auditing usage, resource provisioning and deprovisioning, and movement of users between groups as their organizational roles change. This usually is done from a central dashboard which helps organize the effort and reduce errors. For a small company, this is not a large task but for major organizations with thousands of user roles, this is a huge process. A best practice uses a single sign-on linked to the central directory service. This is particularly important for organizations using SaaS cloud models. In other cloud models, e.g. PaaS and IaaS, identity governance becomes murkier because many resources have special roles and related privileges. Third party tools help with tricky issues and standardize approaches. Figure 7.4 illustrates how relationships between business roles and organizational policies determine user access levels.

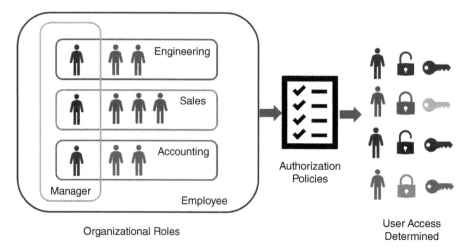

Figure 7.4 Organizational roles and policies determine cloud resource access.

In development environments, a best practice approach develops internal standards for account assignment and use. This includes standard approaches for creating accounts, assigning resources, and removing accounts. Like other objects in the cloud, zombie identities can lead to security problems and provide potential doors into resources for those with malicious intentions. Deprecated identities that have not been removed may allow former employees or others to enter an organization's resources.

Other considerations for governance activities include auditing usage, event logging and analysis, and using analytics to search for unusual usage patterns or unexpected changes to permissions or resources. Audit trails that collect information for the entire usage chain, from user login to accessing resources to logging out, should be put into place. In short, governance plans are a key to successful IAM practices in cloud settings.

IAM Considerations for Developers

Developers now view identity management in ways better suited to cloud environments (Lonea et al. 2013). One view comes from experiences in web-based environments. When a user logs into a secure website, they have access to pages, data, and other resources not visible or accessible by non-authorized users. Think of Facebook for a moment. As a non-logged in user, you can see items like business pages and public profiles. As a logged in user, you see your homepage, feed, friend information, and other resources associated with your account. This is a big difference, of course.

Developers view cloud software implementations in the same way. Applications are *identity-enabled* to ensure security and tie into organizational governance rules. This means that IAM is integrated at the application level and is checked as resources are requested in the application. Developers integrate organizational security and governance features into their code. This means software functions are added into the code. These include:

- New cloud applications integrate with the IAM system to check access rights before each resource access.

- Prior to serving requested data or services, applications perform a series of checks based on data permissions, application permissions, user group, and specific user privileges.
- Data is encrypted according to IAM policies related to the current user, data, and application.
- User access rights are continually checked dynamically as resources are used. This differs from merely passing through a doorway that permits access like the stateless approach used by web servers.

IAM Dynamically: Developer's Approach

A stateless approach for IAM is being built into many cloud software applications. The software uses an API that pulls permission data from a database each time a resource is required. The pattern used in these systems is:

1. **Software requests resources**
2. **Identity retrieved**
3. **Identity validated (usually via API from database)**
4. **Resource access permitted with current identity**
5. **Identity validated and limitations applied**
6. **Resource access permitted subject to permission level and limitations**
7. **Resources released**
8. **Software logs completion**

Identity Provisioning

Identity provisioning is a management challenge for many organizations. This entails securely adding and removing cloud users in timely and effective manners. In addition to those processes, planning must consider the user's potential impact on the system. For instance, do new software licenses need to be added? Does the addition of a user impact processing times, storage needs or network bandwidth? Do systems need downsizing when a user leaves the organization?

In most cloud-based systems, an organization holds a finite number of usable seats or licenses. If projections for use are exceeded, not everyone may be able to log into the system when desired. Therefore, several considerations must go into planning. Included are:

- On average, how many users will be on the system?
- What resources are needed by most users?
- Do enough licenses exist to cover users during critical times and average times of usage?
- Are organizational processes in place that enable business users to inform IT administrators that a new user should be added?
- What resource demand thresholds trigger acquisition of more licenses?
- Likewise, are organizational processes in place that enable business users to inform IT administrators that an existing user should be removed?
- What happens to data regarding removed users? Is it retained or archived?
- Are any specific security measures needed regarding user removal?

Cloud Licenses

Managing cloud software licenses, usually in SaaS environments, goes hand in hand with identity provisioning. Originally, software licenses were purchased on an individual basis and bought just like hardware – as a capital expense up front. In dynamic cloud environments, licensing is more often handled per user as it is consumed. User licensing is paid monthly, often in subscription form. This approach fits well with the pay-as-you-go model used for most other cloud resources as well. Two general licensing models are typical for cloud software vendors. These are:

- Per use
- Per user

Per use licensing monitors either the number of accesses, the amount of time spent in application, or uses other metrics to determine cost. Per user licensing assesses total user numbers to provide a price. Per user approaches may be based on seats, meaning a hard limit on the number of simultaneous users at any given time. Therefore, an organization may have many more accounts on the system than seats, and a license server tracks use to ensure a maximum is not exceeded at any time. Per user licenses also may be based on giving individual users guaranteed access based on login credentials. Subscription prices do not change based on usage patterns, and users always have access no matter how many others are currently in the system.

Some licenses permit users to have multiple copies of the software. So, in addition to using the software in their office, they have permission for home or mobile usage. This model usually applies to situations where a client component of the software is installed on the user's machine. Fully cloud-based SaaS software is location agnostic in many cases (although sensitive software may check user IP addresses and only permit certain usage as part of its security approach).

From an economic perspective, and consistent with justification to move to the cloud, using a licensing model that charges for usage, generally makes most sense. Per-use models ensure companies pay for what is used and are more flexible during peak demand or as employees are added or removed.

Other best practice approaches apply to SaaS software licensing from a business perspective. First, it is important to determine an organization's overall needs and expected usage patterns so all departments using the same software are included in pricing to help with volume discounts. Having a central contact point for software purchases may make sense. This provides flexibility if maximum user numbers exceed licenses. Also, be sure to understand the terms of use. Will scaling cost extra? Ensure the license matches expected usage patterns for the organization.

When acquiring software licenses, an IT administrator should form a good relationship with the vendor's sales team and communicate organizational needs clearly so the correct license can be obtained. Maintaining a good relationship with open communications helps ensure a win-win situation as an organization grows and changes over time.

IAM with Third Party Vendors

Another challenge facing IT administrators deals with multiple cloud vendors. This is particularly common when using SaaS services. For instance, an organization may use

MS-Office 365, Salesforce, SAP, and SAS to accomplish different tasks. By default, most of these services provide users with IDs and passwords. While some companies opt to go with this approach, it is not considered a best practice and can lead to problems. Users may decide they cannot remember multiple different passwords and will reuse the same ones. This leads to weak password protections and provides intruders armed with old passwords from one service, the ability to compromise a different service.

Many organizations use federated ID management (FIM) to ensure users have single sign on for all services. This means the organization selects an ID provider such as Google, Microsoft or Okta. Often, solutions from these vendors are marketed as IDaaS (ID as a service) specifically intended for cloud-based systems.

FIM Versus SSO

Single Sign On (SSO) and Federated ID Management (FIM) solutions have the same goal – users have one set of credentials to access all organizational computing resources in a secure and trustworthy manner. FIM can be considered a form of SSO but the two have key differences. SSO enables users to have a single set of credentials for use in a single organization. FIM expands the definition and provides users with access across different organizations or cloud systems. Not all SSO solutions use FIM but most FIM solutions rely on SSO technologies across domains. SSO is token-based technology which means each user is assigned a token that provides access to systems within the organization. In infrastructures using a federated approach, users interact with an FIM system that provides permission to enter affiliated systems.

FIM systems essentially centralize identity management services using protocols that enable vendors' systems to request permission for use and then receive appropriate information back from the FIM server that grants permission and provides other details about that user's privileges. Multiple enterprises make agreements to allow their subscribers to use the same identification data to obtain access to partner systems without needing to reauthenticate. FIM systems generally use security assertion markup language (SAML) to ensure a wide range of systems can use federated logins. Popular FIM systems include OpenID, OAuth, and Shibboleth. The FIM process uses a procedure like this:

- Users authenticate on their home security domain network.
- After that authentication, users can log into remote, partner networks that use identity federation.
- An application requests authentication information from the user's home authentication server.
- The home server authorizes the user to access the remote application.
- The user is permitted access to desired remote resources.

Shibboleth

Shibboleth is a widely used, open-source FIM often deployed in higher education environments. It allows SSO within organizations and permits federated logins across

---

(Continued)

Figure 7.5 Shibboleth. Source: From Shibboleth, Shibboleth© Internet 2.

(Continued)

enterprises and domains. It also permits sites to protect online resources and ensure privacy of critical data. Figure 7.5 is open source from the Shibboleth website located at: https://www.shibboleth.net

This means the user only authenticates once at his or her home domain. Subsequent accesses to other networks are handled without additional logins.

FIM Benefits

FIM simplifies the IT administrator's role when it comes to tracking users. When a new user joins an organization, a single home network identity is provisioned for them and then they can use partner remote systems. This is better than attempting to add users to many separate services using scripts or manual entry. It also prevents security issues from occurring when a user leaves an organization. One removal from the home domain ensures the user no longer has access to other resources and domains.

Other FIM benefits have economic and convenience motivations. For instance, if an organization partners with a consulting firm, it can form an identity federation to ensure users have access across both firms as needed. When a partnership or project ends, the identity federation can be removed. From a user standpoint, the benefits are apparent. There is no need to remember multiple passwords or user IDs. Updating passwords is easy. IT help desk issues are less likely to occur, and end-users are more likely to have a strong password rather than weak variants of the same one. In short, resource access is streamlined for the user.

FIM Challenges

FIM is not without challenges. An IT administrator must go through the learning curve and ensure the current system is compatible with the federated one. If not, migration might be required. At a minimum, an agent, which is a piece of software from the FIM service provider, must be installed on the local network. Also, an organization incurs costs related to training and, if using commercial services, associated subscription costs.

Another challenge relates to moving from an individual organizational security policy to one considering requirements of all federated members. Companies must understand their own security needs plus those of all organizations they are joining. Perhaps more importantly, an organization must ensure any third party using their resources meets internal security and usage policies. While this sounds straightforward, it can become a tricky task, particularly if it means people from other firms can access your company's internal resources.

User visibility becomes more of an issue. Tracking who accesses each item while maintaining security, privacy and other concerns can be more complex. Often, security audits and logs mean an IT administrator must work with people from outside their firm to uncover usage patterns and needs.

Identity and Access Management Products

As mentioned in prior sections, FIM products, also known as identity management as a service (IDaaS), have spread rapidly across the corporate world and are key elements of ID infrastructure. We covered many reasons for this, but one should be emphasized. Authentication across partner firm software systems is simplified but still ultimately controlled at the local level. This often requires installing a software agent on the local network which permits an IDaaS system to communicate with the local user directory. Once the IDaaS system is satisfied that a login is legitimate, it permits that same ID to use all authorized, connected systems. Local administrators use their own, familiar tools in most cases to manage local user pools.

Several IDaaS providers sell products and services to support cloud computing in this area (Kumar et al. 2018). We already mentioned Shibboleth, but many others exist. We look at a few for a sense of tools available to IT administrators.

Microsoft Azure Active Directory: Azure Active Directory (Azure AD) builds on Microsoft's Active Directory (AD) services to provide multi-tenant, cloud-based identity management services. It focuses on access management, services provided by Microsoft AD, and identity protection across multiple platforms. Azure AD is considered the gold standard of FIM products because of its close integration with AD, Microsoft Exchange, and other Microsoft business products. Unlike some FIM software, Azure AD does its authentication on its own server rather than permitting each organization to authenticate locally and then provide access to federated resources. This means Azure AD offers secure password synchronization and for many companies, which may be exposing their resources, this is reassuring. Microsoft authentication is more trusted than a partner firm's authentication about which they may have little knowledge.

What is Active Directory?

Microsoft's Active Directory (AD) is a Windows OS service used to connect users with network resources. It offers an interface to give IT Administrators access control over network directories, folders, and resources. It controls user access based on security policies set up and organized by an IT administrator.

Google Cloud IAM: Google offers this service to provide cloud identity and access management services. Hence IAM stands for Identity and Access Management. This tool gives administrators the ability to determine which users can view, change, or access various cloud-based resources. It enables security policy development and enforcement. It offers auditing and reporting service capabilities as well. Cloud IAM allows users to access Google-provided services and tools (e.g. Docs, Gmail, Sheets, and so forth) but does not perform user authentication for services outside of Google like Azure AD does. Instead, IT administrators still need to use an authentication service on the local network (such as Microsoft AD) to connect outside of Google.

Okta: This identity management solution works across many emerging major platforms in cloud computing. It uses an organization's on-premise or cloud identity management system and integrates that with its cloud-based tool called *Universal Directory (UD)*. Okta uses an agent running on the local network's domain (which can be virtual). An installation wizard walks the IT administrator through the process of connecting local identities to UD. Okta's Identity Management software can pull user data from sources used by the organization. For instance, if user data is stored in an HR database, this information can be pulled out, modified as needed and fed into another system with different data requirements. So, it becomes a sophisticated tool that enables integrations between many types of systems, both on-premise and in the cloud.

Amazon Cognito: Amazon offers federated identity pools that enable users with identities from various providers to access their resources. Unlike the other tools we discussed, Cognito is intended to help users working in one of Amazon's clouds, often from a mobile platform. The tool allows an IT administrator to create unique identities with customized permissions or to permit users authenticated elsewhere to enter Amazon resources. Cognito pulls user profile information into directories stored on AWS so when they move to another service, the same information can be used. Depending on user needs, the stored information can be customized and synchronized as users move from service to service. Cognito relies on *SAML (Security Assertion Markup Language)*, an open source standard for exchanging authentication and authorization information between identity providers and service providers. It provides information to service providers so secure decisions regarding resource access can be made.

Identity Management Standards

In the prior section we mentioned how Amazon Cognito uses SAML, an open source standard for exchanging authentication information to help manage federated service logins by users. In addition to SAML, several other standards are used in authorization and authentication systems. The most popular of these are OpenID and OAuth2. Having standards permits SaaS system developers ensure users can move in and out without needing to repetitively enter their credentials. This is especially useful in situations where IT administrators want to create an end-user experience where users feel as though they are on a single unified system and can move effortlessly from organizational resource to resource. It also means that these systems do not need to store system passwords because they receive proof of identity from a trusted source like Google or Facebook (see Figure 7.6 for an example).

SAML: This was the original open authentication standard from about 2001. As mentioned earlier, it stands for Security Assertion Markup Language and is used for authentication and authorization. SAML uses the concept of a user, called a principal, wishing to access a resource. In this case, the service provider owns the resource and an identity provider is custodian of the principal's credentials and other identity information. When the principal asks for authorization to use the resource, the resource provider sends an XML (or JSON) message to the identity provider asking for the user's credentials and other needed information. The identity provider returns an *assertion* which contains specific information about that user. The resource provider then can act on the information and allow (or deny) the user access to the resources. See Figure 7.7 for an example.

Figure 7.6 Logging into a third-party system using proof of identity from a trusted source. Source: Google LLC.

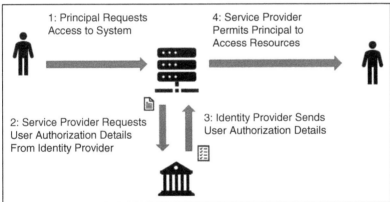

Figure 7.7 Authentication process typically used with SAML.

To illustrate further, we examine a SAML use case. The MyBusiness Corporation uses a SaaS cloud model for its applications which include Salesforce, SAS, and BambooHR. MyBusiness's IT administrator uses a federated approach to allow a single sign on (SSO) that makes users' experiences seamless and feel more integrated. Thalman Brunston is the HR manager at MyBusiness. He comes in at about 8: 00 a.m. every morning and logs into the system via an Intranet portal. After logging in, he clicks on the Salesforce link. Salesforce receives an SAML assertion from MyBusiness, who was established as a trusted identity provider previously. Thalman is logged onto Salesforce automatically and his session is populated with provided data.

OpenID: This open standard is widely used for authentication by large Internet organizations including Google, Yahoo, PayPal, Amazon.com, Flickr, Blogger and over one million web sites requiring a login. OpenID is straightforward: a user gets an OpenID account through an identity provider (which can be Google, Flickr or many others). The user can then log into any OpenID websites which are called the relying party. The OpenID standard ensures the user's details are communicated between the identity provider and the relying party. So, the exchange works as follows:

1) The user obtains an OpenID account which is owned by the user (e.g. http://myOpenID.openid.com) from an OpenID identity provider (e.g. Google).
2) The user goes to a different Website, perhaps Blogger and sees the "sign in with Google" option and clicks on that button or link.
3) The Blogger website contacts Google's OpenID service and receives an association handle.
4) The user is forwarded to the Google login page where he enters his OpenID and password.
5) Google validates the credentials and redirects the user back to Blogger with a validated token.
6) Blogger trusts the Google login and token and permits the user access to the website.

OAuth2 (Open Authorization): This is another open standard used for authentication and authorization with Internet-based systems (although it is not limited to this). OAuth2 is token-based which means after a user enters a correct user ID and password, they are given a token which enables them to access a resource. The token remains valid for a pre-determined time and may also be accepted by other resources. System authentication is managed behind the scenes, so the user does not need to re-enter their credentials again. The token is useful to save time and frustration, but also can be used to enter systems where the user does not want to disclose their password or user ID. These sites only see the token and not user credentials. OAuth2 uses what is called *delegated access*. This means clients (which are the applications) access resources and take actions for the *user*, without the user ever sharing their personal information including ID and password with the application.

Here is an example. A user creates an account with an OAuth2 provider (for instance Twitter). They see a button that asks if they want to import contacts from Facebook. They click on that choice and are redirected to their Facebook account where they log in. After login, Facebook asks if they want to share their friend list. They say "yes" and are sent back to Twitter with a token (invisible to them) that gives permission to get the list from Facebook. Twitter never sees their Facebook credentials but by using the token, can access their resources on their behalf. This process is called *flow*. So, OAuth2 gives an application permission to perform a task or use a resource at the user's request. Figure 7.8 provides a depiction of the differences and similarities between OpenID and OAuth2.

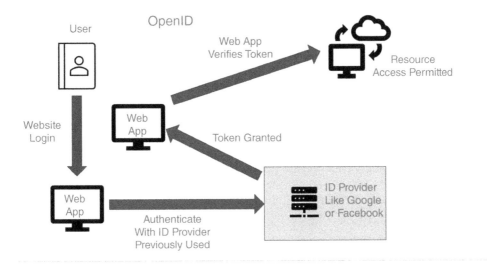

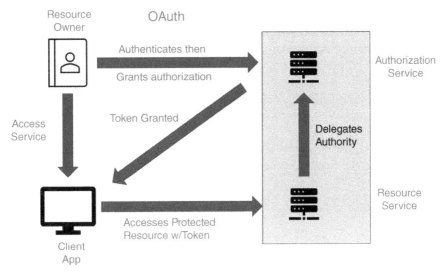

Figure 7.8 OAuth permits an application to access user resources while OpenID uses one service's ID to log into another. Source: Perhelion, OAuth permits an application to access user resources, 2011.

Summary

Ensuring resource security is key to cloud technology success. Users must be authenticated and authorized to enter and use organizational systems in ways that ensure system protection. Mission critical data and the need to connect systems across geographically distributed regions require IT specialists to develop reliable systems to ensure online resource security. Without a doubt, safety concerns change when firms move IT services into the cloud. This means that security is no longer only an on-premise activity. Instead, a strong partnership between cloud provider and cloud client is required. Information and services are housed on a combination of virtual and physical servers accessible via Internet and other networks. This means new concerns including data thieves, malicious hackers, malware, ransomware, and other issues must be considered.

This chapter focused on tools, procedures and concepts required to keep cloud computing resources and services safe. We reviewed concerns and vulnerabilities for each cloud model with SaaS, PaaS, and IaaS; and looked at where cloud clients' and providers' responsibilities start and end. We also looked at current practices in identity management and reviewed how open standards for authorization and authentication can be leveraged by cloud users to provide seamless and secure use of cloud resources. In this discussion we covered federated identity management (FIM) and how identity management as a service (IDaaS) products are used by cloud providers and third-party vendors to make access more secure for organizations. In short, identity management offers several excellent outcomes for IT administrators. Among these are:

- *Improvement in cloud user productivity*: User sign-in is simplified and the need to manage multiple passwords is eliminated. Access rights are quickly and efficiently updated as user needs and requirements change. New users are quickly added and employees leaving are readily removed from all organizational systems.
- *Improved affiliated user services*: An organization's stakeholders including customers and partners benefit from efficient and fast access to appropriate resources. User groups help manage resources only accessible to permitted classes of users. This ensures data and application security.
- *Help desk demands are reduced*: IT staff members experience fewer requests for help with logins, passwords and other aspects of identity management when federated processes are implemented.
- *Cost savings*: FIM and IDaaS provide cost savings in staff demands and resource costs. Many processes can be automated removing costs due to errors and problem resolution.

References

Hashizume, K., Rosado, D.G., Fernández-Medina, E., and Fernandez, E.B. (2013). An analysis of security issues for cloud computing. *Journal of Internet Services and Applications* 4 (1): 5.

Kumar, P.R., Raj, P.H., and Jelciana, P. (2018). Exploring data security issues and solutions in cloud computing. *Procedia Computer Science* 125: 691–697.

Lonea, A.M., Tianfield, H., and Popescu, D.E. (2013). Identity management for cloud computing. In: *New Concepts and Applications in Soft Computing*, 175–199. Berlin, Heidelberg: Springer.

Bibliography

Amoud, M. and Roudiès, O. (2016). A systematic review of security in cloud computing. In: *Proceedings of the Second International Afro-European Conference for Industrial Advancement AECIA 2015*, 69–81. Cham: Springer.

Bamiah, M.A. and Brohi, S.N. (2011). Seven deadly threats and vulnerabilities in cloud computing. *International Journal of Advanced engineering sciences and technologies* 9 (1): 87–90.

Cameron, A. and Williamson, G. (2020). Introduction to IAM architecture. *IDPro Body of Knowledge* 1 (2).

Grobauer, B., Walloschek, T., and Stocker, E. (2010). Understanding cloud computing vulnerabilities. *IEEE Security & privacy* 9 (2): 50–57.

Kuppinger, M. (2014). Cloud user and access management. *KuppingerCole Leadership Compass Report* 70969.

Shere, R., Srivastava, S., and Pateriya, R.K. (2017, October). A review of federated identity management of OpenStack cloud. In: *2017 International Conference on Recent Innovations in Signal processing and Embedded Systems (RISE)*, 516–520. IEEE.

Suryateja, P.S. (2018). Threats and vulnerabilities of cloud computing: A review. *International Journal of Computer Sciences and Engineering* 6 (3): 297–302.

Website Bibliography

Identity Management

Amazon Cognito: https://aws.amazon.com/cognito

Azure Active Directory: https://azure.microsoft.com/en-us/services/active-directory

Google Cloud IAM: https://cloud.google.com/iam

Okta: https://www.okta.com

Shibboleth: https://www.shibboleth.net

8

What Is Cloud Governance?

Governance in cloud computing revolves around setting up sound structures that enable administrators to make good decisions regarding consistent system performance and user accountability. Governance starts with guiding principles that shape how organizational computing is conducted and results in policies related to services. Good governance sets the tone for an organization's users and the seriousness with which they take policies that ensure the security and safety of systems and resources. Good policies ensure applications and data are maintained safely. But, even with the best intentions, problems can emerge. Good governance minimizes problems and provides a sound framework to consistently and transparently deal with ones that do occur in a manner that best serves the interests of the organization.

Cloud governance is a natural extension of the larger area of IT governance and in most organizations, no distinction is made between the two. Cloud computing is where most organizational IT infrastructure has moved, and governance activities have followed. This chapter provides an overview of IT governance with a specific focus on cloud computing. We discuss keeping resources, particularly data, secure and then finish with discussions of auditing and encryption key management considerations.

IT Governance Overview

Information technology governance or *IT governance* is a term describing the portion of over-all corporate governance dedicated to oversight of a firm's information technology activities, systems, policies, and uses. Its primary focus covers three general areas:

- Oversight of IT assets. In other words, ensuring that infrastructure, systems, and uses are compliant with organizational policies and relevant regulations.
- Ensuring IT assets and systems are properly controlled and maintained.
- Ensuring IT assets provide value and support organizational strategies.

In more general terms, IT governance is assured through transparency of the IT function and the oversight of this area by board-level executives. One function of IT governance ensures that key organizational decisions regarding IT receive careful consideration

Cloud Technologies: An Overview of Cloud Computing Technologies for Managers, First Edition. Roger McHaney.
© 2021 John Wiley & Sons Ltd. Published 2021 by John Wiley & Sons Ltd.
Companion website: www.wiley.com/go/mchaney/cloudtechnologies

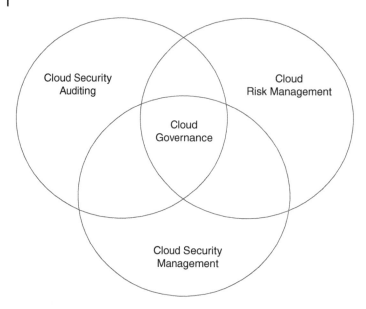

Figure 8.1 Cloud governance occupies a critical junction of risk management, security practices, and auditing.

at a high level so critical needs flow from top level planners to those implementing infrastructure. This requires two-way communication. IT experts must ensure board members understand organizational systems, opportunities, and limitations; and the board needs to ensure IT administrators are aware of organizational needs, future direction, and the role IT plays in revenue generation. This creates a structure for communication and oversight. Figure 8.1 illustrates how IT governance (in the context of cloud computing) fits in the center of three organizational areas: risk management, security, and auditing to ensure the systems operate as expected.

Traditionally, many corporate boards would leave the IT function in the hands of experts. The stereotype view of IT was to locate it in the basement of a building, out of sight, where it would magically develop and roll out systems that worked as needed. IT managers were gurus that provided new capabilities to the business and in many ways determined the organization's direction and decisions.

Those days, which truly never may have existed, certainly are gone now. IT administrators should be included in corporate level planning where they can provide valuable insight and input into the development of organizational mission and strategic plans. It is particularly important for an organization to have a chief information officer (CIO) or equivalent and for this person to participate with other senior management members in high level planning. The CIO manages a highly complex infrastructure of hardware, software, data, networks, and applications that enable an organization to achieve its goals. Strategic directions and goals must consider what is realistic in terms of IT capabilities and be aware of future trends and opportunities. Therefore, a primary feature of IT governance establishes organizational relationships that build a nexus between business operations and IT.

IT Governance Boards

In most cases, particularly for larger organizations or corporations, IT governance involves establishing a board. This board can be independently created, or be a committee formed from the larger board of directors, or even be a function of the existing board of directors. Often, this hinges on firm size and the importance of IT to an organization's mission. Generally, the board includes both business representatives and at least one IT manager (e.g. CIO).

What Is a CIO?

The *Chief Information Officer* or *CIO* is usually the most senior technology position in an organization. The CIO traditionally managed the IT function in relative isolation and ensured the IT function ran behind the scenes. IT capabilities determined what a firm could accomplish. Today, that has changed. CIOs are strategic thinkers charged with IT transparency; organizational and outside collaboration; and, enhancing business value. CIOs must understand their business and see where technology is moving. They are strategic business leaders that drive change and transformation. Their role includes both overseeing operations and planning how IT can help enhance business opportunities leading to growth. CIOs are connectors and must ensure IT connects to business operations, vendors, customers, and other stakeholders.

Larger corporations often create an IT governance committee formed of independent directors with experience in business and strong connections to technology. They may not be IT executives but should be IT-savvy individuals with experience in IT oversight from prior organizational roles. The best boards often include at least one member who is a current or former CIO.

IT Governance Frameworks

As mentioned earlier, IT governance ensures a business's IT investments and operations align with the strategic plans and operations of a business. Usually a formal structure, in the form of a board, provides the mechanism to carry out the functions of IT governance. Most corporations see IT governance as required oversight based on laws, business practices, and as a measure to protect their firm from deception, fraud, and high-profile data theft issues.

In most countries, businesses must comply with a wide range of regulations that protect personally identifiable data in ways that include keeping confidential information safe, ensuring secure data storage, overseeing backup and recovery practices, and ensuring investors are confident that sound practices are in use.

To ease stakeholder worries, most organizations rely on existing IT governance frameworks to implement their processes. Several popular frameworks and standards exist; and these have been adopted by thousands of organizations. Among these are: COBIT, ITIL, AS 8015-2005, ISO/IEC 38500:2008, CMMI and FAIR. These frameworks offer a road map for IT oversight and suggest standard methods to evaluate IT performance and effectiveness. Overall, the frameworks provide evidence of legal and regulatory compliance regarding IT practices. The next sections look at several frameworks in more detail.

COBIT 2019

Many experts consider COBIT the gold standard IT governance framework. It offers a process for implementing IT governance within an organization from a business objective perspective (De Haes et al. 2020). An organization called ISACA (Information Systems Audit and Control Association) created and maintains the standards. They provide a range of training programs, consulting references, and other services for IT oversight and auditing. ISACA provides an overview of COBIT and says that it has been created to[1]:

- Maintain high-quality information to support business decisions.
- Achieve strategic goals and realize business benefits through the effective and innovative use of IT.
- Achieve operational excellence through reliable, efficient application of technology.
- Maintain IT-related risk at an acceptable level.
- Optimize the cost of IT services and technology.
- Support compliance with relevant laws, regulations, contractual agreements and policies.

In practice, COBIT (currently implemented as COBIT 19) comprises several components. Among these are: framework, process descriptions, control objectives, management guidelines, and maturity models. A short description of each follows:

Framework: The framework is an overall approach used to implement standardized best practices regarding IT infrastructure. The primary goal of the framework is to align IT implementations with overall business strategies and goals. This mechanism encourages communication between IT administrators, other executives, and business managers.

Process descriptions: COBIT includes non-technical descriptions of best practice processes to ensure IT is not hidden behind a shroud of technobabble. This ensures all company executives can talk about IT and communicate with IT specialists in a common language.

Control objectives: These are requirements for IT practice put into language meaningful to non-technical executives. These develop and provide improvement goals for all IT processes that align with a business's specific practices.

Management guidelines: COBIT provides best practices for business objectives and describes how responsibility for various tasks should be assigned within an organization. It also offers benchmarks and measurement suggestions with a focus on organizational integration (if other frameworks also are in use).

Maturity models: COBIT offers insight into a firm's IT lifecycle and how processes and approaches should change over time. This is a growth-based perspective used to prevent problems associated with phases in a business's maturity.

ISACA provides helpful support with COBIT and its implementation. Training courses and a recommended implementation path ensure an organization develops its IT governance to a high level. IT executives and others involved in auditing and governance can become certified in COBIT. Figure 8.2 shows the main areas of emphasis for COBIT implementation.

---

1 Source of these components is: https://cobitonline.isaca.org/about

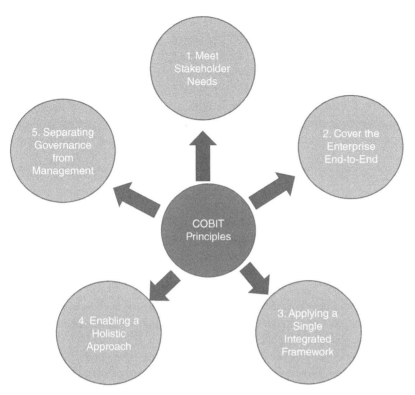

Figure 8.2 Principles of COBIT that ensure IT governance within a firm. Source: - Principles of COBIT that ensure IT governance within a firm, 2016.

Overall, COBIT provides knowledge for senior management so they gain insight into ways technology can be used and how it aligns with organizational goals (Chandra and Wella 2017). Organizational IT maps directly into points within COBIT helping identify missing IT components. In fact, COBIT provides a reference model with over 40 IT processes useful in organizational infrastructure. COBIT's best practices carefully describe each process, inputs/outputs, key activities within the processes, and measurements to determine effectiveness. Other tools include ROI measures and best practices to integrate COBIT with other standards and frameworks perhaps already used within an organization.

ITIL (Information Technology Infrastructure Library)

This framework originally was developed within the British government in the 1980s to provide best practices information for its IT functions. The Central Computer and Telecommunications Agency originally released more than 30 books with information from industry best practices. Those volumes were condensed into five books that are supported currently. The intention for Information Technology Infrastructure Library (ITIL) was to help government operations standardize technology use and ensure IT supported core functions and needs. The framework uses a service-oriented perspective and has evolved into a general framework meant to support business operations.

Organizations (both governmental units and businesses) can send employees to ITIL certification training. The certification is aimed at IT management and staff. A key outcome of the training ensures all IT personnel share a common language to facilitate issue resolution. ITIL is a good choice as a governance framework for organizations with service-oriented IT needs. Many organizations view this framework's service desk best practice guidelines as the gold standard in this area. But that is a small part of what ITIL offers. Overall, it provides best practice approaches for management in five general areas: strategy, design, transition, operation, and continuous improvement. Figure 8.3 offers a view of the subcomponents within each area.

AS 8015-2015

This standard came about in response to the Australian Government's finding, after the dot-com bubble crash in the early 2000s, that poor corporate and information governance caused many business failures. So, the Australian Standard for Corporate Governance of Information and Communication Technology was developed and published. It provides principles, common language, and best practice models for corporate ICT (information and communications technology) governance. This standard moves away from a perspective that focused on management systems and processes, and instead provides advice that separates management practices from governance.

As stated within the standard, "Corporate Governance of Information and Communication Technology (ICT) is the system by which the current and future use of ICT is directed and controlled. It involves evaluating and directing the plans for the use of ICT to support the organization and monitoring this use to achieve plans. It includes the strategy and policies for using ICT within an organization."[2]

Within that context, AS 8015 breaks down its framework into standards for directors (e.g. IT governance board members) and those with daily responsibility for business and IT operations within an organization. The framework provides a governance model, advice for monitoring and evaluating ICT, a common vocabulary, and *Six Principles for Good Governance of ICT* which include[3]:

1. Clearly delineate responsibilities for ICT.
2. Carefully plan ICT to best support the organization.
3. Ensure the acquisition of ICT is valid.
4. Ensure implemented ICT performs as expected, if not better, when needed.
5. Verify that ICT conforms to a set of formal rules.
6. Ensure ICT respects human factors.

While many organizations still use AS 8015, it has been superseded by other standards such as ISO/IEC 38500:2008.

---

2 From AS 8015-2005 Standard.
3 This information from: *da Cruz, M. (2006). "10: AS 8015-2005 - Australian Standard for Corporate Governance of ICT". In van Bon, J.; Verheijen, T. Frameworks for IT Management. Van Haren Publishing. pp. 95–102. ISBN 9789077212905. Retrieved 23 June 2016.*

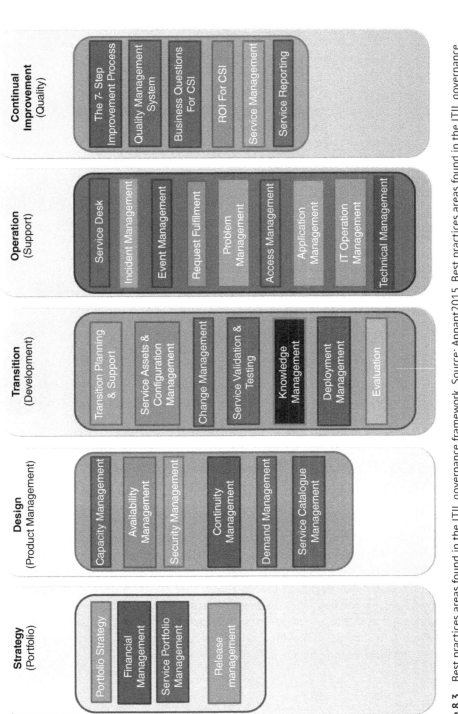

Figure 8.3 Best practices areas found in the ITIL governance framework. Source: Annant2015, Best practices areas found in the ITIL governance framework, 2013.

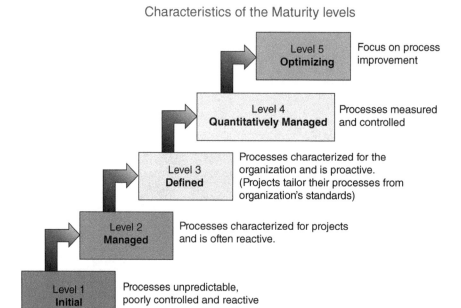

Figure 8.4 Five stages in the CMMI model. Source: NASA, 5 stages in the CMMI model.

ISO/IEC 38500:2015
Like its predecessors, ISO/IEC 38500:2015 offers guiding principles for organizations. The standards provide guidance for director level executives including governance board members and those responsible for daily business operations. This standard was influenced heavily by AS 8105 and recommends clear delineation between management and governance to ensure oversight of sensitive issues. This standard also recognizes that ICT processes may be controlled either on-premise or by service providers (e.g. cloud vendors). This is significant because of its relevance to modern business uses of the cloud.

CMMI
This acronym, derived from the words: *Capability Maturity Model Integration*, was developed by the Software Engineering Institute (SEI). CMMI determines where on a scale from 1 to 5 an organization currently operates in terms of IT performance level, quality, and profitability. The current state of organizational IT drives process improvements required to get to the next level. Several best practices and measurements are recommended. The CMMI has limited applicability in cloud environments because vendors often supply mature solutions. However, it is useful to provide an overall picture of a firm for governance boards. Figure 8.4 provides details of the five CMMI stages.

FAIR
Factor Analysis of Information Risk or *FAIR* is a newer model frequently used for IT governance. This model focuses on organizational risk. It was developed in the era of cybersecurity and cloud computing, so these aspects of risk informed the approach's development by

The FAIR Institute. This non-profit organization of risk specialists, IT cybersecurity leaders and other managers, and executives state their mission as:

> Establish and promote information risk management best practices that empower risk professionals to collaborate with their business partners on achieving the right balance between protecting the organization and running the business.[4]

FAIR provides business leaders with insight suggesting at-risk aspects of their organization and then offers standards and best practices to mitigate potential loss; and to better understand, measure, and report factors that influence risk. The central tenant of FAIR is a taxonomy or classification of potential risk factors that affect each other. These factors are weighted to provide probabilities for data loss. By reducing risk factors, the overall potential for data loss is reduced. FAIR uses probabilities because risk is uncertain and many outside events can exert influence.

FAIR views risk as the probability of loss tied to an asset. FAIR suggests an asset may add value and liability to a firm. For instance, maintaining information about a potential customer is valuable to a firm. On the other hand, if that data is stolen, the outcome could greatly damage the firm. Therefore, it costs money to protect the data asset. FAIR seeks to help a business understand and minimize the liabilities from a cost and risk perspective. Based on this perspective, FAIR's six potential loss categories for an organization include:

1. *Productivity*: Firms can no longer produce goods or offer services due to a disruptive event.
2. *Response*: Resource loss due to disruptive event.
3. *Replacement*: Cost to replace an asset damaged during a disruptive event.
4. *Fines and judgments*: Legal costs due to a disruptive event.
5. *Competitive advantage*: Missed opportunity for new business due to disruptive event.
6. *Reputation*: Cost due to reputation being tarnished following a disruptive event.

FAIR Categories for Value and Liability

1. *Criticality*: How big of an impact does the loss have on productivity?
2. *Cost*: What is the cost to replace a lost asset?
3. *Sensitivity*: How does the loss impact the organization in terms of cost related to embarrassment, competitive advantage, legal/regulatory, and other areas.

FAIR also provides insight into *threat agents*. Threat agents are divided into communities that share common characteristics and are likely to impact assets in similar ways. Threat agents can take the following actions regarding assets:

- *Access*: Read without permission.
- *Misuse*: Use asset without permission or for reasons other than intended.
- *Disclose*: Provide data to others without permission.

---

4 Mission from: https://www.fairinstitute.org/mission

- *Modify*: Change without permission.
- *Deny access*: Block others from use.

The value of an asset, the type of action taken against it, and the loss categories determine the overall impact on a firm. Overall, FAIR helps organizations understand potential for loss and offers factors to preemptively stop problems.

IT Governance in the Cloud

As we have seen, IT governance provides oversight within an organization to ensure IT operations, including systems, data, services, software, and hardware, are implemented in ways that ensure the organization is protected and usage aligns with organizational goals and strategies. Governance should be separate from management and ensure best practices are incorporated in cost-effective ways, and that required regulations are followed. Governance operates similarly in the cloud but may have a different focus since the challenges are unique.

Cloud governance must take cloud services specifically into account. Depending on the cloud model used (e.g. SaaS, PaaS, or IaaS), governance activities have different foci. For instance, with SaaS, most of the security and other infrastructure will be carried out by the cloud vendor. Therefore, the governance committee will be concerned with the SLA and whether it adequately protects the firm in the event of problems. The governance committee will be interested in the indemnity clause and other measures used by the vendor. It is likely the committee will scrutinize the SLA to ensure compliance with regulatory requirements and consider cost elements of the agreement. With PaaS and IaaS, the activities of the governance committee will more closely resemble those dealing with on-premise concerns.

The distributed nature of the cloud makes governance difficult in many ways. In PaaS and IaaS environments, the governance committee will want to determine who is responsible for security and privacy; and how the responsibilities are divided and enforced. Compliance issues become a concern as will other aspects of the SLA such as performance and so forth. The governance committee may wish to investigate the cloud provider or review the SLA prior to it being signed.

Just like IT governance, cloud governance policies are created and reviewed by a committee of experts that have IT and business experience. Having a cloud expert would also be beneficial. The group develops policies that ensure organizational operations are safe, compliant, cost-effective, and forward thinking.

An area of concern in cloud governance relates to data visibility and ensuring corporate data remains within the organization's control for the long term. In SaaS environments particularly, departments, groups, or individuals within an organization may independently roll out their own applications. This could result in political conflict within the organization regarding logins, data storage, and so forth. Governance can ensure these issues do not occur and that policies are followed to safeguard the long-term viability of IT and data resources. This area falls into the category of *shadow IT*. Governance committees often develop strategies to deal with this complicated area within cloud computing. Later in this chapter we will look at shadow IT in more detail.

Other areas of concern for cloud governance are web applications, social media, and mobile deployment. Often, IT departments have little control over these operations, and control mechanisms are non-existent. Cloud governance policies should describe exactly who controls these activities; and, how and when data is updated and managed. It may not make sense for IT departments to control how social media and web sites are updated; or to have input into the content being broadcast. However, governance needs to ensure a close working relationship between organizational content providers and those responsible for ensuring the systems operate with a high level of uptime, are protected from adverse events, and maintain data security, privacy, and safety. It is also important that authorization and authentication are managed consistently with the rest of the organization. Losing access to data due to employee turnover is a major concern in these areas.

Choosing a Governance Framework

We reviewed several governance frameworks and suggested extra considerations for cloud implementations. All IT governance frameworks were developed to ensure IT departments and technology-related organizational activities function well and have taken major risk factors into consideration. We saw how governance is conducted by a committee that is composed of experts in business with IT or cloud experience and are separate from organizational management to provide an outside view of the bigger picture.

So, how does a governance committee decide what framework should be used to guide oversight? Many organizations currently use COBIT when risk management is a major concern. ITIL is aimed primarily at service organizations. FAIR provides cutting-edge guidance when cybersecurity and data protection are major concerns. There is no definitive answer for selecting a governance framework but rather, it depends on business mission and operational model, corporate culture, industry risks, and other factors. Some organizations adopt portions of multiple frameworks. COBIT encourages this to ensure coverage that works best. For example, many organizations will use COBIT as a general framework and then integrate ISO/IEC 38500:2015 standards within relevant areas where it makes more sense.

After investigating various frameworks, the governance committee may recommend an approach using one or more frameworks. The committee must ensure that organizational officers buy into the approach and are committed to working within the new framework.

Cloud Risk Factors Related to Governance

IT governance in cloud computing seeks to minimize risks and ensure an organization is protected from a variety of potential threats to its well-being. To effectively discharge this duty, the governance committee must consider potential risks that differ from on-premise implementations. Primary differences we have mentioned previously include the cloud provider and SLA. The IT governance committee should carefully consider the terms of the agreement with cloud vendors and how valuable corporate data and process knowledge will be handled, preserved, and safety ensured should something happen with the vendor. In general, the following areas should be considered in cloud governance:

- *Vendor and SLA risks*: Ensuring vendor agreements and roles regarding all aspects of system use and safety are covered.

- *Data security risks*: Ensure data is safely stored, confidential data is protected, and privacy is maintained as needed.
- *Data and system recovery risks*: Ensure data and systems can be recovered in the event of disaster or unexpected adverse events.
- *User risks*: Ensure user authentication and authorization in cloud systems are adequately implemented, particularly if federated systems are used.
- *Organizational information risks*: Ensure organizational intellectual property is maintained and kept safe.
- *Compliance risks*: Ensure regulations, laws, and corporate policies are followed, particularly in multi-vendor environments.
- *Audit risks*: Ensure audits are conducted regularly and audit trails and data visibility are maintained in cloud-based systems.
- *SLA-related system performance risks*: Ensure cloud-systems can operate as expected regarding availability, performance, and critical usage times.
- *Interoperability risks*: Ensure cloud-based systems integrate and work together.
- *Legal risks*: Ensure indemnity clauses are appropriate and legal protections are in place regarding SLAs and other vendor contracts.
- *Financial risks*: Ensure billing and resource usage are properly implemented and controls are in place to avoid paying for unnecessary resources.

Risk management is a key area of governance. The frameworks that we covered provide a starting point and means for organizing and managing governance activities. But the bottom line is that cloud computing introduces a new element–vendors manage key operations for an organization and need oversight to ensure the best decisions are made to minimize potential risks. Open communication channels and relationships between IT administration, business managers, governance committee members and key vendors can help ensure this critical area accomplishes its intended purpose.

IT Audit Committees

Often, IT governance boards create an audit subcommittee. This committee may be chaired by a governance board member and will oversee evaluation and examination of organizational IT policies, infrastructure, usage, and procedures. Audit activities are conducted in a variety of ways. Among these are: (i) external auditors are hired to conduct the audit; (ii) internal auditors are employed by the organization and work with the audit subcommittee and others; and (iii) auditors come from an external regulatory body to perform audit activities.

In general, audits ensure IT operates according to policies, but are *not* meant to test compliance. Instead, IT auditors examine whether an organization's "relevant systems or business processes for achieving and monitoring compliance are effective. IT auditors also assess the design effectiveness of the rules – whether they are suitably designed or sufficient in scope to properly mitigate the target risk or meet the intended objective."[5]

---

5 Quoted from: Tommie Singleton in the ISACA Journal, Volume 6, 2014, "IS Audit Basics: The Core of IT Auditing," found at: https://www.isaca.org/Journal/archives/2014/Volume-6/Pages/The-Core-of-IT-Auditing.aspx

This means the auditors look at current controls and whether these ensure compliance. If not, then IT administrators must find better ways of implementing systems, security, or whatever area is in question.

IT Auditor

In general, an IT auditor evaluates whether IT systems and controls are adequate to ensure effective compliance with organizational and broader regulatory policies. The auditor does not so much test controls but determines from a comprehensive perspective whether the controls adequately ensure compliance. Internal IT auditors often work with external auditors reporting to the IT governance committee or other external bodies. In cloud environments, IT auditors often work with cloud vendors to ensure their policies, controls, and practices adequately meet internal needs for organizational entities which consume the cloud services.

IT auditors may conduct specific system testing or may work with developers to recommend testing for controls. Further, IT auditors may develop reports and present findings to the IT audit subcommittee and others. These reports may include mandatory changes or updates, recommended changes or updates, and suggestions for policy changes. The auditors may conduct research to understand regulations or an organization's contractual obligations, and then provide guidance to the organization regarding whether current systems are adequate or if changes are necessary.

Some IT auditors are specialized in an industry. For instance, in the United States, higher education entities have FERPA requirements to consider and health care providers must consider HIPPA regulations.

IT Auditor General Duties

- Evaluate IT systems, cloud vendor practices, and processes that secure organizational data.
- Evaluate risks to organizational information and intellectual property assets; recommend approaches to minimize those risks.
- Ensure management and departmental processes related to IT are compliant with laws, policies, regulations, and standards.
- Determine inefficiencies in IT systems, cloud processes, and implementations of IT.

Overall, IT auditors ensure compliance failures do not occur. They seek to understand why an organization might have issues in an area and often use their findings to dig deeper and find an underlying root cause. IT auditors should have experience in IT systems and truly understand both big-picture operations and many details that ensure the system works smoothly. They are part detective, part system integrator, and part business expert. A good auditor is worth their weight in gold to a firm.

IT Controls

IT controls mitigate known risks. In cloud environments, having appropriate controls is very important. Like accounting controls, an organization puts these into place to reduce risks that naturally occur in a process. IT controls can be *preventive, detective,* or *corrective.* Most IT operations seek preventive controls. This means problems are stopped before they occur. Detective controls provide notification that a problem has occurred, and then corrective controls help "fix" the problem. Of course, detective and corrective controls should be closely aligned so when an issue occurs, it can be quickly and effectively remediated.

In its basic form, an IT control usually is a policy or procedure that provides assurance that IT-related usage operates as intended. This means it complies with regulations, laws, policies, and so forth. Controls keep data and systems secure and reliable.

The best controls are seamlessly built into system use procedures, so they are transparent to users and do not add overhead actions. Controls such as these are *internal controls.* However, in cloud systems, this could be a misnomer because the control might be running outside the firm. Therefore, it might be better to think of them as *integrated controls.*

IT best practices generally focus on four main control categories. These are: *manual controls, IT dependent manual controls, application controls,* and *general controls.* Manual controls historically are used in most systems. These include things such as having a permission form signed by a supervisor, signing an Internet use policy after receiving training, or signing a travel reimbursement form indicating that you agree with its accuracy. Manual controls usually are outside the IT system but may use technology in their implementation (e.g. a use policy might be provided and signed online by the employee).

IT dependent controls automate manual controls and provide some system integration. For example, an IT dependent control may provide a list of all people that have logged into a system in the past month and send it to a supervisor for review. Another example might be a report that is generated to list everyone with access to a database but has not used it in the last month. This automatic report may be sent to a manager for further action. In general, IT dependent controls rely on some automation (e.g. report generation) and some manual intervention (e.g. manager reviewing report and taking further action).

When most people think of IT controls, it is application controls that come to mind. There are many ways to implement these controls. They include simple examples such as login and password policies, and account lockouts after three failed attempts. Other examples are more complex and include automation and orchestration activities such as performing security updates on software across a firm. Any configuration settings or filters put on an input field in an application are considered application controls. Any problem that can be prevented this way falls into this category. Often, IT auditors provide advice to design teams and help identify controls that should be built into a system.

The final category, general controls, ensure correct access to resources, govern change management, and make sure physical security has been implemented. This means the right people have access to the right resources (e.g. ensure Active Directory groups are correctly associated with the correct resources). Regarding change management, controls ensure changes are authorized, documented, tested, approved, and that users are trained prior to having access and so forth. And, of course physical security involves a review of the way IT resources are kept safe in offices, rooms, and in cloud environments.

Sarbanes-Oxley (e.g. SOX Section 404) U.S. Securities and Exchange Commission (SEC) (2019)

Sarbanes-Oxley is a U.S. federal law that regulates financial aspects of a public firm with parallels in many countries. Its requirements impact IT audits since IT systems and cloud computing contribute to financial risk. In general, SOX includes requirements to:

- Evaluate any IT systems, cloud vendor practices, processes or procedures used to manage data related to finances.
- Evaluate information risk and the security of intellectual property assets.
- Evaluate organizational IT processes to ensure compliance with standards, laws, regulations, industry standards and internal policies.
- Find inefficiencies that may impact financial practices including those in cloud processes and IT implementations.

The COBIT framework, that we covered earlier, often is used to navigate SOX Section 404 compliance (U.S. Securities and Exchange Commission (SEC), 2019).

We have barely scratched the surface of IT controls. This is an entire field of study that extends beyond the scope of this book. It is important to remember that controls are critical to an organization and those that partner with cloud vendors must consider implementation, use, and maintenance of appropriate controls. Cloud vendors can provide best practice information to make controls more manageable and useful but new issues can arise. IT administrators, IT auditors, and IT governance boards should work together to ensure controls contribute in a meaningful way to safety and risk mitigation. Just because a control can be put into place, does not always mean that it should be. For instance, if a control costs more to implement than replacement of the non-critical asset being protected, it would not make sense to put it into place. Controls can be viewed in terms of costs and benefits. A control costs money to identify, conceptualize, design, implement, and manage. Other costs may include the impact on a process, employee training, and operational inefficiencies. IT auditors seek a balance between costs and benefit to a firm. Assets that are critical must be safeguarded with controls. Those with little value do not need the same level of protection. Due diligence of an auditor helps determine whether a control is necessary or not. Figure 8.5 provides a hierarchy of controls to illustrate different categories that may need to be considered.

End-User Controls

An area of concern in most firms relates to the growing area of end-user computing. In today's environment, people have access to a variety of mobile devices, social media, and cloud-based software subscriptions. For organizational IT administrators and IT auditors, this represents a potential nightmare. Data, information, and organizational intellectual property can leak out in ways never possible in the past. For instance, an employee can quickly load a spreadsheet of company data up to their private Dropbox account, intending to use it during an upcoming meeting. Later, when they are employed by the organization's competitor, that spreadsheet is still in their private Dropbox.

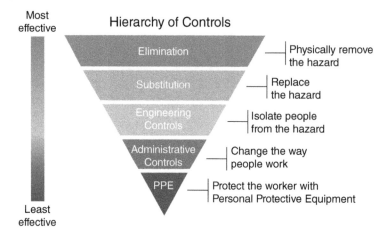

Figure 8.5 National Institute for Occupational Safety and Health hierarchy of controls. Source: National Institute for Occupational Safety and Health Centers for Disease Control and Prevention.

It may seem as though little can be done to curb these behaviors but that is not always the case. Many organizations use a combination of policies, training, and software tools to prevent data or intellectual property from leaking out.

One area of concern involves spreadsheets or personal databases created and used by individuals, specifically those involved in handling financial data and therefore are subject to regulation and SOX 404 assessment by auditors. IT controls often are difficult to implement or even non-existent at this level. Spreadsheets have become so powerful that large datasets, complex calculations, and organizational intelligence can be embedded at the personal level. This increases the potential for errors, fraud, or misuse outside the oversight of the governance committee. In cloud environments, spreadsheets easily can be stored on private accounts or on shared drives giving access to people that otherwise would not be granted permission to restricted data. A wide range of public cloud spreadsheet tools exist and if employees use, for instance, Google Sheets to develop a spreadsheet, it becomes easy to share that item with others outside the organization, or with those not having official privileges to access the data. So, organization-wide policies and training should be developed to ensure these important, flexible tools are used in responsible and transparent ways (Wallace et al. 2011).

Developing appropriate policies is a starting point for ensuring end-user controls exist within cloud environments. If employees understand organizational expectations and the reasons behind the policies, they will be more likely to exercise caution regarding how they handle data and sensitive intellectual property. Training reinforces policies and helps emphasize potential consequences associated with inadvertent leakage of sensitive data or intellectual property. Staff members should be apprised of ways to ensure safe use of end-user software such as passwords on spreadsheets, encrypting sensitive information, using secure corporate drives rather than personal cloud subscriptions, and so forth. Giving employees an easy-to-use range of alternatives to their personal tool set helps set expectations and ensure they can perform their job duties effectively.

Organizations can use cloud drives configured with a file syncing service to back up employee spreadsheets in ways that make their jobs easier. Development of a central storage area for end-user applications gives employees a convenient method of storage and helps ensure security. In addition to training, providing easy-to-use alternatives to private cloud subscriptions, and suggested behavior changes can have a big impact.

Organizations may wish to use other, formal measures to ensure security. Included are:

Content-limiting filters: An organization can work with its ISP to open a subset of web sites accessible by employees in their work environments. Many governmental agencies take this approach in secure settings. Filters can consider regulations, laws and corporate concerns and limit social media sites or other places where data leakage might occur. While this is effective, without providing training or justification for the filtering, employees might feel minimized and not trusted by their firm. Therefore, careful implementation focusing on the reasons for these measures should be used.

Network filtering: Organizations can implement filtering at the network level within their on-premise network or work with their cloud provider to implement filtering over virtual networks. Filtering can block data loss and ensure files do not leave the organization's secure areas.

There is no 100% guarantee that organizational data will not leak outside the secure perimeter meant to safeguard it from misuse. An organization and its IT auditors should perform an end-user risk-analysis to identify ways this can happen. Responsibility for control over end-user application use, personal use of public clouds, and mobile devices is a shared responsibility. IT administrators, management, and business users must buy into policies that safeguard knowledge and data and commit to using tools and approaches that minimize risk and ensure compliance with regulations. IT administrators must supply secure alternatives for personal use and the users must do the rest. Figure 8.6 looks what might happen otherwise.

Shadow IT

In some organizations, corporate end-user applications have been eclipsed by individual, departmental, or project team use of cloud-based applications from outside the formal IT infrastructure. This phenomenon is known as *shadow IT*. The term, shadow IT, sometimes is used to refer to all unauthorized IT use in a firm but we will use this to mean information technology applications managed by an organizational unit without formal knowledge of the IT department. This could be as simple as a spreadsheet used in a department by its members or as complicated as a major SaaS cloud-based application used to perform most of a department's work duties.

Shadow IT has grown incredibly in the past few years with some experts suggesting more that 35% of IT spending in organizations takes place outside the formal IT structure. This has happened because of the wide array and high quality of applications available, the ubiquitous nature of mobile computing, the pressure to work from home after hours, social media, availability of low cost SaaS systems, slow response from corporate IT groups regarding departmental needs, and new collaboration tools. Many departments have their own

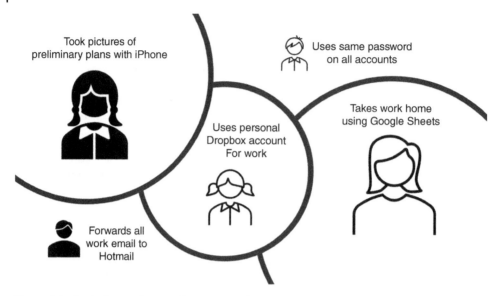

Figure 8.6 Lack of controls on end-user computing in cloud environments.

independent IT operations run by their own staff members due to perceptions that support is focused and faster.

Shadow IT is both a challenge and a benefit. For example, employees are freed from corporate systems that may be slower or less specific to their needs. Corporate help desks may have fewer demands since support comes from within the department. On the other hand, data leakage, lack of controls, and data loss potential all exist. Likewise, organizational knowledge that could be shared may not be known outside the department. Outside regulations, audits and policies may not be conducted correctly. If something goes wrong and a data breach occurs, the organization is still responsible, and further, the CIO may be held accountable even if they had no idea the shadow system existed.

Therefore, an IT administrator may have to block shadow IT applications. This could mean ensuring organizational firewalls prevent unauthorized applications from running on organizational computers. But this might cause employees to run applications on their own hardware using a hotspot. Working with the people using the shadow IT is sometimes the best approach.

Performing risk assessment can determine if some shadow applications are low risk. Perhaps these are okay for employees to use. If an application is high risk, perhaps the organization can find a way to incorporate departmental needs into its infrastructure or secure the application. Working with end-users rather than in opposition to them builds trust and helps ensure surprises are not going to occur down the road.

Acceptable Risk

Risk can never be eliminated from an organization. This is particularly true in cloud computing environments. Rather than eliminate risk altogether, IT auditors talk about acceptable risk. This means considering cost and probability of occurrence, then developing a smart solution with reasonable protection against critical problems.

IT auditors seek to understand and offer recommendations to reduce risk but, in some situations, acceptable risk levels remain. Thorough auditors identify these areas and recommend contingency plans in event of a worst-case scenario. Another consideration for IT auditors is to balance risk in one area with controls in another. For instance, consider a particularly high cost associated with a low probability event that could be quite damaging to a firm should it occur. A good auditor can recommend controls in other areas of the organization that could be implemented at a lower cost but still be effective and prevent major problems. This holistic view of an organization emerges with an auditor's experience and ability to communicate with organizational managers and IT administrators.

SOA Governance

One last area of governance we will examine is called *service-oriented architecture* or *SOA* governance. SOA governance has been in existence for a long time and bears many similarities to cloud-computing governance. In fact, many experts view cloud services governance as a direct extension of SOA governance with new considerations such as multitenancy and elasticity.

SOA governance is put into place to ensure service quality, predictability, visibility, and cost-effective performance within an organization. It also ensures that policies, laws, and regulations are followed by an organization. So far, this sounds a lot like all forms of governance that we have covered. The difference is in the way SOA governance is composed. It has three primary components which follow:

- *SOA registry*: This is a catalog listing SOA services available. It can be used internally or as a tool to enable development of business partnerships
- *SOA policy*: Principles used to ensure services do not conflict and to ensure implementation follows good design, custom relationships, and compliant practices
- *SOA testing*: A regular schedule of audits, tests, and performance metrics used to ensure operation of SOA. It intends to ensure SOA solutions are working correctly in a secure, cost-effective way. It also ensures regular system updates.

Figure 8.7 looks at a generalized SOA implementation. Notice how security and governance are balanced against operational considerations. Cloud computing practices often are informed by SOA best practices.

Ensuring Secure Cloud Data

A primary function of governance, particularly when overseeing cloud environments, is to ensure data security. Without a doubt, data breaches, losses, and corruption are the biggest, and most publicly damaging risk factors for many organizations. Recent news stories about loss of customer data have resulted in massive financial damage to organizations and the demise of others. Even with a comprehensive SLA and indemnity clause, an organization

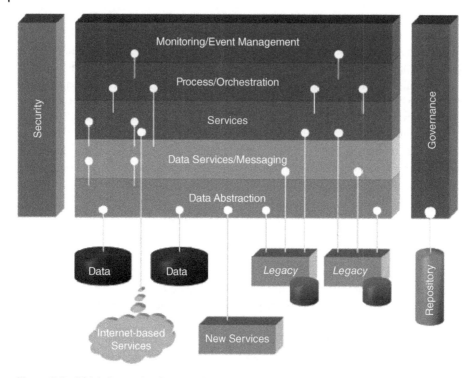

Figure 8.7 SOA informs cloud computing governance and security. Source: Loïc Corbasson, SOA informs cloud computing governance and security.

cannot afford to be the subject of bad publicity in this area. For this reason, many governance boards take extra measures to ensure data security. Among considerations are[6]:

1. SLAs contain clear language that verify cloud providers take strong measures to ensure cloud data are maintained in accordance with their security guidelines and compliance policies. A full set of requirements should be clear and available to governance board members.
2. That cloud providers ensure relevant government regulations which may include HIPPA, FERPA, and the EU's General Data Protection Regulation (GDPR) are carefully implemented to ensure personally identifiable data is maintained securely and privately.
3. Performance tools monitor network, database, and applications to detect suspicious access or unexpected movement of data, particularly if large amounts are moved without prior knowledge.
4. Use of data encryption tools (which we will cover in more detail later in this chapter).
5. Use of multi-factor authentication approaches.
6. Use of IP blocking tools for critical applications and data access.

---

6 These 10 items are derived from suggestions provided by the Software Engineering Institute at Carnegie Mellon University (https://www.sei.cmu.edu).

7. Use of appropriate firewalls which may include virtual as well as more traditional physical firewall devices.
8. Use of sound access key management practices.
9. Use of tokenization for authorization and access to resources when appropriate.
10. Use of CASB to ensure overall security of cloud resources.

Of course, the best practice for cloud safety is to maintain due diligence. Keeping up to date on new security strategies and tools is important, as are being informed about recent data breaches and issues that resulted in these happening. Ensuring data is backed up and having secure APIs for critical data access are also important considerations.

Cloud Provider Data Safety Measures

Encryption is a widely used approach to ensure data safety. As mentioned in earlier chapters of this book, cloud providers often recommend encrypting data prior to transit and for data stored or at rest on their servers. Cloud vendors may offer options regarding encryption keys. Either they will maintain and control them, which of course means they do have access to the cloud customer's data on their service. Or, if that is unacceptable to the governance board, then the keys can be maintained by the cloud client directly. That means the cloud client has complete control over who can access and use their data.

Other methods also ensure data safety. Among these are a process called *sharding*. This approach breaks files into small chunks and then encrypts each separately. Each chunk is stored in a different location on the cloud provider's servers. That way, if an intruder happened to break through normal defenses and decipher the encryption scheme, they would only have random chunks of data, meaningless without knowing how to piece them together.

Other cloud vendors provide links to online files used for a preview mode to help minimize unnecessary downloads. The reasoning is that extra downloads could result in unwanted files being sent to trash on computers where someone may have access without permission. By providing links, a preview is offered ensuring the file is the one desired by the authorized user prior to download.

Cloud Encryption

Encryption is the most effective way of securing data. It scrambles files, data, or other stored items according to a complex pattern based on a key. Only someone possessing that key can unscrambled the contents in a reasonable amount of time. Cloud providers offer different approaches to encryption depending on a client's security needs. One is to encrypt data prior to transmission to their cloud where it remains encrypted until the client retrieves it. This is called *end-to-end encryption*. The second approach is to only encrypt the most sensitive data like passwords or customer credit cards numbers. This is called *limited encryption*. In some cases, the encryption is done after the cloud provider receives it and is completely managed by the vendor. In other scenarios, cloud clients may wish to encrypt data themselves without involving the cloud vendor. This ensures no one working for the cloud vendor can access their data in an unencrypted form.

> ### 256 Bit Encryption
>
> Encryption comes in many forms. One of these, 256-bit encryption, is commonly used for data being transmitted from one server to another. The 256-bit part refers to the encryption key's length. To break this encryption, a hacker would need to try up to 2^{256} combinations before absolutely finding the correct outcome. This would be a herculean task for the most powerful computers.

The benefits of encryption are obvious: data are readable only by intended audiences or users that have access to decryption keys. When data is encrypted, it is useless to someone that may have stolen it and does not possess the decryption key. Cloud encryption also is important for compliance reasons. Governmental and other regulations stipulate data privacy and security (for example HIPAA, FERPA, and SOX laws in the U.S.). Encryption ensures this compliance.

While encryption increases data security, it also adds overhead and slows down data storage and retrieval processes. So, an IT auditor may recommend that only sensitive or critical organizational data be encrypted. Like any security feature, costs and benefits should be considered. Since encryption requires more bandwidth, some cloud customers encrypt their own data on-premise prior to transmission to the cloud to reduce overhead and preserve bandwidth. Other vendors may offer cost savings alternatives meant to reduce processing consumption and bandwidth such as data redaction or obfuscation of just key fields. Proprietary algorithms and software can repopulate the confidential fields at the time the data is needed.

Cloud encryption schemes require building trust with vendors. For this reason, many cloud vendors work with their clients to develop secure solutions to data and file storage. Several complicating factors exist that open potential security risks which need to be addressed. For instance, how will the encryption key be managed? Vendors will often pass the encryption keys to their clients so that data can be decrypted after transmission back to them. It is important to remember that ultimately, security risks must be managed by the cloud client and need more than just vendor contracts, controls, and IT audits.

A set of policies should be developed by an organization and its governance committee to ensure data is protected. These requirements can be shared with the cloud vendor to ensure the data is encrypted as desired. First, an organization should decide what data needs encryption. For example, if data being stored is used for training purposes and contains video and sample data sets, it may not need encryption at all. Perhaps only account passwords and user IDs need to be stored securely. On the other hand, customer health data may be sensitive and need to have end-to-end encryption to ensure its privacy. For data that is highly sensitive, the organization may choose to encrypt it on-premise and maintain the encryption key themselves on their own internal server. That protects against attacks that occur within the cloud environment. Another consideration relates to compliance. If an organization is subject to government oversight or stores data that could be used in a criminal investigation, a subpoena or other request might come to the cloud provider. If they hold the key, they could supply the data. The cloud client may rather have the request come to them and can ensure this happens by being the sole holder of the decryption key. Managing encryption keys can be complex and requires secure systems. Although this textbook

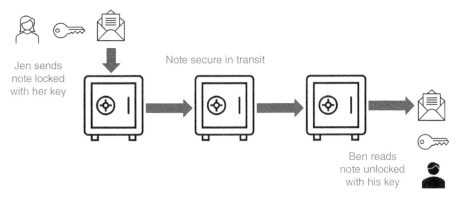

Figure 8.8 Simplified illustration of symmetric key encryption with Jen and Ben. Source: Some icons courtesy of www.publicdomainpictures.net.

is about cloud computing, it is important to describe how encryption works with a little more detail. While many complex and proprietary encryption solutions exist, most can be categorized into two broad groups (Chandra et al. 2014):

Symmetric key algorithms: These provide either identical or very similar keys for both encryption and decryption operations.
Asymmetric key algorithms: Unrelated, completely different keys are used for encryption and decryption operations. A very commonly used approach to encryption, public key cryptography, falls into this category.

Symmetric Key Encryption

Symmetric key encryption assumes the same key can be used to both encrypt and decrypt data that is being stored or shared. So, to illustrate, consider the following example:

Ben and Jen want to keep their messages secret until the moment when they announce their engagement to their respective families. They currently have a long-distance relationship, and this complicates their correspondence. The last time they met in person, they purchased a lock box that had two identical keys (since neither uses social media nor modern technologies!). Each kept a key when their weekend rendezvous was over. Ben took the lock box with him to Seattle where he currently lives on a boat. He wrote Jen a heartfelt message, locked it into the box, and mailed it to Jen who currently lives in a high-rise apartment overlooking the Chicago River. She received the box via courier and used her key to unlock the box and read the delightful message. Unable to contain her passion, she dashed out a response, locked it into the box and sent it back. Of course, Ben will open the box and read the message with trembling hands when it arrives later in the week (see Figure 8.8).

Symmetric-key algorithms generally use a stream or a block approach. The main difference being the way the message is broken up prior to encrypting. In a stream cipher, each character is encrypted one by one to provide a ciphertext stream. In a block cipher, which is considered more secure, blocks of text (often 64-bit chunks) are encrypted simultaneously. This requires more computing power but creates a more secure result. Many algorithms

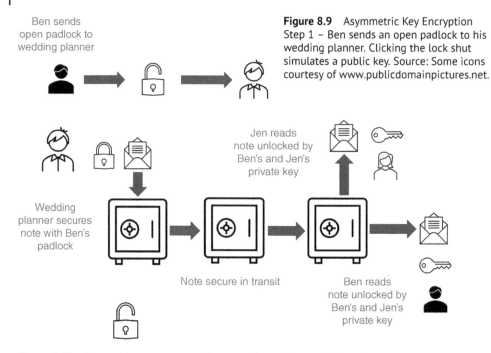

Ben sends open padlock to wedding planner

Jen reads note unlocked by Ben's and Jen's private key

Wedding planner secures note with Ben's padlock

Note secure in transit

Ben reads note unlocked by Ben's and Jen's private key

Figure 8.9 Asymmetric Key Encryption Step 1 – Ben sends an open padlock to his wedding planner. Clicking the lock shut simulates a public key. Source: Some icons courtesy of www.publicdomainpictures.net.

Figure 8.10 Ben and Jen use asymmetric encryption to keep wedding planner messages private. Source: Some icons courtesy of www.publicdomainpictures.net.

exist for symmetric-key encryption and these are beyond the scope of this text. A few of the more popular ones are: RC4, Twofish, AES, and Blowfish.

Asymmetric Key Encryption

Asymmetric key encryption uses a different approach to encryption. This process is intended to protect the key as much as the message content. In an asymmetric key system, anyone can lock a message using a public key but only those with the correct private key can unlock it and view the content. We will return to the Jen and Ben example to illustrate:

As the date for announcing their wedding drew closer, Ben and Jen needed to send secret messages to their wedding planner. They wanted to be sure the planner could see what was needed but did not want them to accidently obtain access to one of their steamy, private messages. They realized it was time to start using an asymmetric approach to encrypt their messages. To do this, Ben first sent an open padlock to their wedding planner (see Figure 8.9) but kept this private key for himself (Jen also has a copy of this private key!).

The wedding planner then wrote their message and used the public method to lock it (which was to click the lock shut!). The message went to Ben and he used his copy of the private key to open it (Figure 8.10). Later, the wedding planner sent an open copy of their padlock so Ben could send a private response using the same method. Figure 8.11 illustrates.

In asymmetric key encryption, keys are never transmitted. This adds security to communication and helps prevent an intermediary party from acquiring the keys. It also ensures a variety of people can send messages to a specified receiver without being able to read

Figure 8.11 Wedding planner sends open padlock to Ben for response.

Wedding Planner sends open padlock to Ben for future message

each other's messages since the process uses a public key for encryption and a completely different key for decryption. The message receiver has both a public and private key but only sends the public key for encryption. In the case of our example, this was simulated by sending an open padlock. The public method for locking it was to click the lock shut.

Other Encryption Methods

Of course, symmetric and asymmetric encryption are just two of many encryption technologies available. As cloud computing becomes more sophisticated and threats to security grow, there is little doubt that technology will mature in this area.

Another form of encryption sometimes used in cloud computing is called data masking. This approach replaces sensitive data with fictional data that can be translated back at key times using algorithms built into software applications. This enables the data to be used for testing, data analytics, and other activities without exposing the contents unnecessarily. Often, when data is shared with third party developers, academics, or others, this approach will be used. That way, work can be completed but no regulations are violated. This is particularly important in environments subject to scrutiny such as those using student or health data.

Secure Sockets Layer (SSL)

Another commonly used form of encryption in cloud computing is called secure sockets layer (SSL) protocol. This is a widely used approach for securing communications between a browser and web server. It is intended to reduce the chances that someone will intercept communications. SSL is transparent to users and this helps ensure its wide use. Since SSL is a protocol, used to provide a secure channel between two applications or devices, it can be used for other purposes such as for an internal network. When it is used with web-based communication, the website's address is changed to start with HTTPS instead of just HTTP. The "S," as you probably have guessed, stands for "secure."

In general, SSL uses the following pattern for encryption. First, a user types a request for a website into their browser. This usually takes the form: www.google.com or any other website using SSL. Transparent to the user, the browser requests the web server's certificate. The web server sends it to the requesting computer. The browser checks the certificate for authenticity. It does this using a trusted certificate authority. Most web browsers have public keys of these authorities pre-installed. It uses that key to ensure the certificate is valid and checks to be sure it has not expired. The browser verifies that the domain name and usually IP address for the web server match what the trusted authority

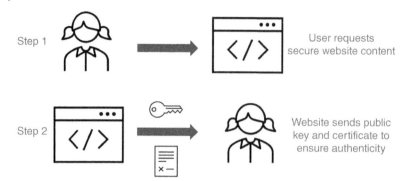

Figure 8.12 First two steps in using SSL protocol for secure web site communication. Source: Google LLC.

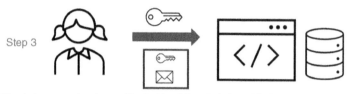

User's browser checks certificate. If valid, website's public key
used to encrypt new random symmetric key sent to server
with encrypted request data

Figure 8.13 Third step in using SSL protocol for secure web site communication. Source: Google LLC.

have listed for its credentials. Once assured, the browser gets the public key of the web server which is also part of the certificate.

> **Certificate Authority**
>
> A certificate authority is an independent organization that issues digital certificates to ensure a web site belongs to the company claiming to represent it. The certificate authority seeks to stop middleman attacks or web site frauds where a site might pose as a legitimate company to obtain private information.

Now the browser knows the web site is legitimate so it generates a shared, symmetric key that can be used for further communication with the web server. This shared key is more efficient than the asymmetric initially sent by the web browser. The browser does, however, use the asymmetric public key of the web server to encrypt the symmetric key for transmission to the server to ensure communication privacy on the rest of the exchange. When the web server receives the browser's transmission, it decrypts the shared symmetric key using its private asymmetric key and then uses the browser's shared symmetric key to decrypt the request data sent along with it. From that point forward, the browser and server exchange data via the shared key. Figures 8.12–8.14 further illustrate the process.

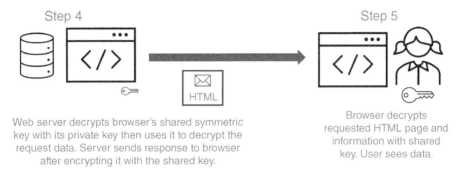

Web server decrypts browser's shared symmetric key with its private key then uses it to decrypt the request data. Server sends response to browser after encrypting it with the shared key.

Browser decrypts requested HTML page and information with shared key. User sees data.

Figure 8.14 Fourth and fifth steps in using SSL protocol for secure web site communication. Source: Google LLC.

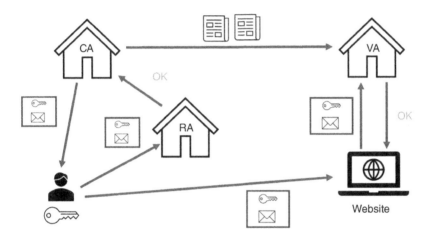

Figure 8.15 Process used by a certificate authority. Source: Chris 🏛 Process used by a certificate authority.

Although we know what a certificate authority does, it helps to get a little more insight into the process used by these important service providers. Figure 8.15 provides an illustration of the process in action. In this example, we see that first, a requestor (which usually is the organization with a website) applies for a certificate, using an asymmetric public key. This goes to the registration authority, represented as RA on the figure. The RA does a due diligence check of the requestor's identity and credentials and notifies the certification authority (on the diagram CA). The CA issues an official certificate. This certificate is available for the requestor until its expiration date. The requestor can use it to digitally sign contracts or place it on a web server as proof of identity and ownership. Each time a digital business contact is made, the certificate, now available through a validation authority (VA on the figure), provides proof of identity for the requestor who needs to ensure the certificate is valid. The top CAs include Comodo, IdenTrust and Symantec. The CAs work with subordinate organizations that are the RAs.

Key Management

As you may have realized by now, managing keys for encryption can be a critical task for an IT administrator. A typical organization will have their own private keys and several shared keys from cloud vendors and other business partners. A few commonsense practices need to be considered. For instance, keys should never be stored with encrypted data. Keys should be backed up and stored in secure locations safe from harm that may befall an organization's IT infrastructure in times of disaster or adverse events. Key security practices should be audited regularly to ensure the latest technologies and threats are understood and put into practice. Keys need to be changed regularly and their expiration dates monitored to avoid unexpected problems.

Some organizations maintain their own keys using on-premise procedures they have developed. Other organizations rely on vendors to provide key management services. It is also possible to use a combination of approaches depending on security requirements, laws, and other factors. In all cases, the governance board and IT audit team should be involved with key management practices.

Cloud computing does not change the basic premise of using keys, but it does add new complications, challenges, and concerns. It also provides innovative and safe practices that vendors have created to ensure safety. Among these practices, cloud vendors have introduced both hardware and software solutions. The hardware approach is known as a *hardware security module* or *HSM*. The software approach is known as a *key management service or KMS*. We will look at both approaches and see how each can be implemented successfully in various cloud computing situations.

HSM: HSMs traditionally are implemented for on-premise key management practices. Using a hardware approach makes security practical for internal networks but adds complexity when a user has moved their IT operations into the cloud. Some cloud vendors have virtualized HSMs for client use, but the risks are greater because most cloud service providers recommend using KMSs.

KMS: As mentioned earlier, this is a software solution for securing and managing an organization's keys. In this approach, keys are maintained in the cloud environment and secured using algorithms and best practice procedures. In most instances, the keys are only accessible to the cloud client. This approach offers several advantages common to all cloud computing: scalability, centralized management, usage monitoring, ability to apply security rules and safeguards, and so forth. However, many experts believe KMSs are not as secure as HSMs, except perhaps when both are cloud-based. But, as we known, HSMs can be hard to use in some situations, for instance, when an organization is geography dispersed or uses multiple clouds and systems. As with HSMs, it is important to keep keys and data separate when using KMSs.

Hybrid Approach: Some cloud vendors have worked to combine HSMs with KMSs for their clients. In this situation, the cloud providers offer a physical HSM device within their cloud data center. The client uses that device for their key storage. This solution provides a higher level of security than the KMS alone but has several drawbacks. First, the keys are located on the provider's hardware and are therefore outside the client's immediate control. Also, if the client uses multiple clouds or has some on-premise computing, the hybrid approach might become difficult to manage. A cloud client may find themselves attempting to manage multiple key systems under this scenario. The keys are in the same environment as the data and this also could create a risk. On the plus side, a solution such as this would work well for a client using a single cloud provider and offer many of the commonly described cloud advantages such as scalability and so forth.

HSM as a Service: Cloud vendors and third-party software companies have combined HSM capabilities with cloud service models to offer a high-level of security in multi-cloud environments. These systems are cloud-neutral and globally available with low latency. This means the keys can be served out to the systems very rapidly. Data and keys are maintained separately to prevent access in event of data breaches. Several products offer HSM as a Service, but these are still being carefully evaluated by cloud users. One example is Equinix SmartKey.

Key Management System Products

Major cloud vendors all offer products and services for key management. We will review what Google, Amazon, and Microsoft currently recommend.

Google KMS: Google currently offers Cloud KMS. This is a software based, hosted KMS aimed to replace on-premise key systems. Its main strengths involve its integration with existing Google's Cloud IAM and its audit logging system. These tools enable keys to be managed at very granular levels. Likewise, permissions, monitoring and alerts are readily available to enable higher levels of security. The key system is automated and elastic, giving users the ability to manage millions of keys. Cloud KMS also enables a user to integrate their key services with the hardware-based Cloud HSM service to ensure compliance with certain laws.

AWS KMS: Amazon offers a key management system that integrates with its other cloud management systems. Like Google's solutions, AWS KMS integrates with other Amazon services to permit assigning permissions, tracking use, setting up alerts, monitoring and performing analysis.

Azure Key Vault: Microsoft offers a service called Azure Key Vault to encrypt keys and safeguard passwords. It has the capability to import keys from HSMs or to generate new keys. It ensures that key contents are not seen or extractable by Microsoft or its employees at the data centers. Keys can be protected using permissions, audits, event management, and threat detection algorithms.

Summary

Chapter 8 provided an overview of ways organizations seek to ensure their IT operations use good practices that make financial sense as well as ensure data and applications are safe from adverse events and comply with laws and regulations. Generally, a committee composed of experienced business and IT experts offer guidance to IT administrators and oversee operations. This group is called a governance committee.

Cloud Governance starts with guiding principles used to shape the approach to organizational computing. The result is a set of policies related to services. Good governance is responsible for setting a tone for users and ensuring policies cover requirements appropriately. Although governance sets the tone, IT administration must implement recommendations. IT auditors ensure policies match requirements and the systems in place accomplish what is needed. They provide recommendations used to strengthen the firm's computing infrastructure and work with cloud vendors to ensure best practices are in place and computing infrastructure is up to date.

This chapter specifically focused on cloud governance concerns as organizations move more deeply into cloud computing. Auditors in cloud environments need to focus on new challenges such as data safety, shadow IT, end-user controls, data encryption, key management, and others. This chapter provided an overview of these concerns and offered insight into roles for governance and auditing.

References

Chandra, K. and Wella, W. (2017). Measuring operational management information technology: COBIT 5.0 and capability level. *IJNMT (International Journal of New Media Technology)* 4 (1): 37–41.

Chandra, S., Paira, S., Alam, S.S., and Sanyal, G. (2014). A comparative survey of symmetric and asymmetric key cryptography. In: *2014 International Conference on Electronics, Communication and Computational Engineering (ICECCE)*, 83–93. IEEE.

De Haes, S., Van Grembergen, W., Joshi, A., and Huygh, T. (2020). COBIT as a framework for enterprise governance of IT. In: *Enterprise Governance of Information Technology*, 125–162. Cham: Springer.

U.S. Securities and Exchange Commission (SEC) (2019). Sarbanes-Oxley Section 404 – guide for small business. Washington DC. https://www.sec.gov/info/smallbus/404guide.pdf.

Wallace, L., Lin, H., and Cefaratti, M.A. (2011). Information security and Sarbanes-Oxley compliance: an exploratory study. *Journal of Information Systems* 25 (1): 185–211.

Further Reading

Agrawal, M. and Mishra, P. (2012). A comparative survey on symmetric key encryption techniques. *International Journal on Computer Science and Engineering* 4 (5): 877–882.

Al-Ruithe, M. and Benkhelifa, E. (2017). A conceptual framework for cloud data governance-driven decision making. In: *2017 International Conference on the Frontiers and Advances in Data Science (FADS)*, 1–6. IEEE.

Bounagui, Y., Mezrioui, A., and Hafiddi, H. (2019). Toward a unified framework for cloud computing governance: An approach for evaluating and integrating IT management and governance models. *Computer Standards & Interfaces* 62: 98–118.

Huang, X. and Chen, R. (2018). A survey of key management service in cloud. In: *2018 IEEE 9th International Conference on Software Engineering and Service Science (ICSESS)*, 916–919. IEEE.

Rebollo, O., Mellado, D., Fernández-Medina, E., and Mouratidis, H. (2015). Empirical evaluation of a cloud computing information security governance framework. *Information and Software Technology* 58: 44–57.

Steuperaert, D. (2019). Cobit 2019: A significant update. *EDPACS* 59 (1): 14–18.

Svatá, V. (2019). COBIT 2019: Should we care? In: *2019 9th International Conference on Advanced Computer Information Technologies (ACIT)*, 329–332. IEEE.

Yu, W., Yu, P., Wang, J. et al. (2018). Protecting your own private key in cloud: Security, scalability and performance. In: *2018 IEEE Conference on Communications and Network Security (CNS)*, 1–2. IEEE.

9

What Other Services Run in the Cloud?

The cloud offers many options to organizations ranging from replacing its entire infrastructure to just supplementing key software applications. In addition, it is possible to use the cloud for specific purposes to replace various IT functions in an organization. This chapter focuses on odds and ends that fit into this category. We first look at DevOps to discuss how the cloud is changing the way that applications are developed and deployed. We then look at microservices, a software design approach that leverages cloud flexibility in small increments. After that we cover topics related to DevOps such as cloud database applications, analytics services, Hadoop, and open source private cloud software. This chapter reveals ways that IT specialist roles are altered by new cloud technologies and innovative ways of solving problems.

DevOps

You may remember from an earlier chapter that DevOps is an integration between an organization's software development practices and its IT operations. We briefly described how the cloud enables developers to work closely with those running their software to automate the process of creating and deploying new or updated applications. The key to DevOps is collaboration: software developers and IT teams build, test, and release software in reliable and rapid ways that focus on teamwork and integration. DevOps breaks down functional silos to create a new culture that enhances an organization's ability to solve problems and deploy solutions leveraging expertise from two related areas. The cloud provides a platform for this to occur and ensures agile software development practices are integrated into outcomes that make orchestration and automation a seamless part of the process.

DevOps appeared in the corporate world around 2007 when it became apparent that the cloud would permanently alter the ways software development and deployment would take place. The gulf between developers and operations people in IT fields became an impediment to new solutions that had appeared. For instance, many software applications developed in-house were now available for purchase. Some software developers became configuration specialists, and this required closer integration with operations. From an IT administrative perspective, the separation between these areas was no longer acceptable and new practices were put into place. But, perhaps more importantly, in many organizations, the DevOps movement was not a top-down mandate but a grassroots realization

Cloud Technologies: An Overview of Cloud Computing Technologies for Managers, First Edition. Roger McHaney.
© 2021 John Wiley & Sons Ltd. Published 2021 by John Wiley & Sons Ltd.
Companion website: www.wiley.com/go/mchaney/cloudtechnologies

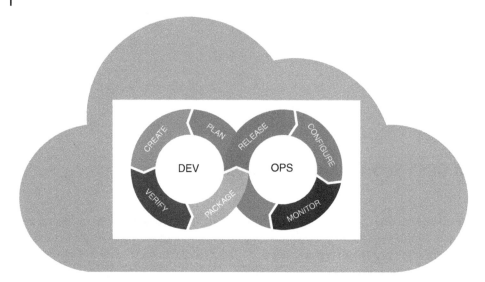

Figure 9.1 An illustration of the DevOps process – integration between development and operations in a cloud environment. Source: Adapted from Kharmagy, An illustration of the DevOps process.

that more efficient ways of working existed. So, many new solutions occurred within organizations organically. Figure 9.1 provides a view of DevOps in a cloud environment.

DevOps Ingredients

For DevOps to work in an organization, several key ingredients need to be in place. Although the ideas seem to be common sense in many cases, remember that decades of software development and operations culture within organizations had to change due to business demands and the availability of new tools and techniques in the cloud. The primary ingredients are described in the following sections.

Ingredient #1: Communication

While technical expertise is a prerequisite to the DevOps area, without a doubt, communication, and interpersonal skill have moved to the top of the required attributes list. DevOps uses agile methods to create business value from technology solutions. This means that the people doing the work need to communicate with each other effectively; and they must work closely with non-technical colleagues from HR, marketing, sales, production, accounting, finance, training teams, and other internal stakeholders. They also can expect to deal with vendors (e.g. cloud providers), governance teams, client organizations, third party software vendors, and many others. So, more than ever, IT specialists must move from being isolated experts to becoming proactive leaders in the organization. Important changes include moving toward a common language that avoids technical jargon. Specialists must remember that most people in an organization are not familiar with catchphrases, acronyms, or highly technical terms. A common language must be established. Control Objectives for Information and Related Technology (COBIT) and other governance frameworks help with that process. Another danger is expecting too much from email communication. Today's social media tools and communication tools (e.g. texting, collaboration tools, and the like) make a huge difference in keeping team members on the same page, organized, and interacting frequently. Communication breakdowns are prevalent in organizations but can be recognized and mitigated.

Ingredient #2: Collaboration

Collaboration goes hand-in-hand with communication in DevOps. In fact, DevOps emerged because of the need to integrate development and operations in IT. So, at its most basic level, DevOps is about collaboration. Hence, collaboration is its key cultural attribute. Key elements of collaboration include communication, as mentioned above; the need for rapid and formative feedback; and modern toolkits to enable and enhance interaction. Often, collaboration does not come naturally to people. Instead, it can be learned and fine-tuned over time. For a team to develop, people need to trust one another and learn each others' strengths and weaknesses. A good team builds relational links (Warkentin et al. 1997). This means they get to know each other better. Part of collaboration is having clear goals. Teams may not function well if each person has a different agenda. Setting common priorities and managing those in transparent ways helps greatly. This also helps a team develop a sense of identity and work together better. Inclusion of diversity and welcoming points of view informed by different experiences, backgrounds, and cultures also contribute to collaboration. Sensitivity together with recognition that various perspectives exist is a great way to achieve better, well-rounded outcomes. Finally, having a clear and well-defined road map that spells out roles and key objectives ensures each member's contributions fits into the big picture. Having regular meetings, deliverables, checkpoints, and interactions also can help.

Cross Functional Teams in Cloud Computing

In cloud computing, DevOps teams must be cross-functional. Taking this approach offers a number of valuable benefits. Among these are:

- Team members bring their specific area focus to the team.
- Team members learn about other areas and their concerns.
- Knowledge of cloud computing benefits and challenges filters back to each team member's home area.
- Channels for open information and discussions emerge.
- DevOps specialists are moved into the mainstream of an organization.
- Specialists can become motivated by knowledge of area challenges.
- Team robustness is enhanced.

Ingredient #3: Flow

Agile principles help ensure DevOps achieves its goals. An important aspect of this philosophy is *flow*, see Figure 9.2. Flow means value-adding work continually occurs without delay between process steps. Each activity enhances the overall project in the shortest time frame

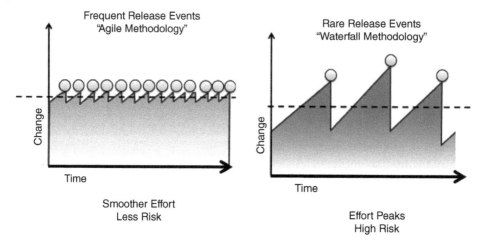

Figure 9.2 Lean cloud approach builds "flow" compared to older design methodologies like the "waterfall method." Source: Christopher Little, Lean cloud approach builds 'flow' compared to older design methodologies like the 'waterfall' method.

possible. DevOps integrates development with operations, as you already know. Therefore, flow moves forward (e.g. development to production) but incorporates feedback loops from production directly back to the developers in constructive, integrative ways. Operations activities must include business user feedback and support functions. Likewise, developers engage in customer support activities. This is not their primary duty but occasionally being on-call helps solve end-user issues and enhances understanding of the impact of development concepts in the real-world.

Ingredient #4: Continuous Improvement

Flow operates seamlessly with continuous improvement (Swaminathan and Jain 2012). The feedback loop that moves from end-users back to developers underlies the basic premise of DevOps: continuously finding problems or weaknesses and working to resolve them with long-term, forward-thinking solutions. The idea of continuous improvement has a long history in organizational settings, starting with early quality gurus like Edward Deming (Walton 1988) and Phil Crosby (1979). Its integration into software development modernizes the approach to building systems that capture user requirements and respond quickly to changing needs in a high-quality manner (Lewis 2017). Continuous improvement emphasizes the importance of moving quickly to improve so business opportunities are not missed. DevOps teams must have a culture that desires to become better and solve problems. Many experts recommend a formal problem-solving methodology to aid in this process. Better-known examples include: Define, Measure, Analyze, Improve, and Control (DMAIC) from Six Sigma, Theory of Inventive Problem-Solving (TRIZ), and Kepner-Tregoe. We will examine DMAIC and TRIZ in more detail in the next section to provide insight into how cloud computing and continuous improvement have a natural synergy.

Ingredient #5: Lean Computing

Lean approaches to development and operations coincide with continuous improvement. Cloud environments lend themselves to lean because they are highly responsive to change in organizational demands and enable organizations to ensure resource usage matches resource demand (e.g. elasticity). A lean approach means "less is more" and cloud resources should only be deployed to the level needed and no more. DevOps teams ensure this happens through monitoring tools and usage demand reports available on cloud platforms. Lean computing ensures everything used has a purpose and any item not adding value to the organization is removed. Learning is key to good implementation. Therefore, design choices are made as late as possible to eliminate rework as more knowledge is gained. The following principles comprise lean computing:

- Eliminate Waste
- Continually Learn
- Empower Teams
- Ensure Integrity and Transparency
- Decide as Late as Possible
- Deliver Rapidly
- View Systems Holistically

Another lean principle is self-organization. While leadership within a group will emerge, perhaps on both technical and managerial levels, it comes about because of the group's internal needs and project demands. Team leadership revolves around understanding and guiding the group in needed directions, learning, teaching, coaching, and continual improvement.

Ingredient #6: Tool Kit

DevOps works in a new environment leveraging cloud capabilities. Collaboration tools enhance DevOps group needs. The best tools include collaboration, interaction, and organizational features that enable group members to connect in non-intrusive but sophisticated ways at times that work best. Additionally, cloud-focused DevOps teams require tools to support orchestration and automation. For this reason, most teams rely on a portfolio or DevOps team toolset to support activities. Table 9.1 offers ideas where tools might be applied and provides suggestions. More information about these tools can be found on their websites.

The best tool kits for DevOps teams operate in their environment and offer intuitive interfaces to ensure productivity and reduce problems. One useful tool, with general purpose uses, is Trello. It provides both mobile and desktop interfaces to permit sharing resources, schedules, emails, timelines, alerts, and other artifacts with team members. Both group and individual processes can be organized and maintained. While Trello might be good for many uses, remember that a DevOps team needs support for the entire development and operations effort. The team must deal with cloud-related business and infrastructure issues that require support. One of the most important areas is version control to ensure the correct software is in use by operations and being updated in development.

Ingredient #7: Quality

We have discussed lean approaches, continuous improvement, and agile methods in this section. All comprise a broader concept: quality. In a DevOps environment, everyone must ensure that service delivery is the highest quality. Teams must view the ultimate outcome

Table 9.1 DevOps tool kit.

Tool	Use
GitHub	Version control/source code management
Microsoft Visual Studio	Version control/source code management
Puppet	Configuration management/automation and orchestration
Chef	Configuration management/automation and orchestration
SharePoint	Resource sharing/collaboration
Zoom.us	Collaboration/interaction
Trello	Organization/resource sharing/interaction
Nagios	Monitoring

as a shared responsibility. The term, *collective accountability*, is used in many organizations to describe that a team has a common goal: a healthy IT system for the organization.

Cloud-Based Problem-Solving Approaches

Cloud computing and DevOps teams often require problem-solving techniques. Some problems can be solved intuitively or with a small group using informal methods. In fact, if possible, problems should be solved in a simple, straightforward manner. But often, problems are complex and require a formal approach to ensure that all aspects are considered in a logical and consistent manner. Cloud computing, like any organizational knowledge work, can approach problem-solving with formal approaches like DMAIC, TRIZ, or Kepner-Tregoe. We will examine two of these: DMAIC and TRIZ in more detail.

DMAIC

DMAIC is widely used due to the popularity of Six Sigma, which incorporates this approach to problem-solving. Essentially, DMAIC provides a roadmap useful for organizing quality improvement processes or entire projects. The letters in the acronym "DMAIC" represent the five primary steps used to solve a problem. These are: Define, Measure, Analyze, Improve, and Control. Figure 9.3 illustrates the process and groups the steps into general categories.

The "Define" stage determines the problem and sets project goals. In the Six Sigma world, problem-solvers set numeric or precise goals with specific outcomes. Impacted business processes are included in the definition. Cloud-computing problems should be specifically defined as well.

"Measure" better determines how the problem can be quantified. For instance, in a cloud computing situation imagine the original problem was defined as, "the *Average Disk Queue Length* is above 4 more than 80% of the time. The *Average Disk Queue Length* should be above 2 no more than 5% of the time." In this case, additional measurements would help the team determine and understand why the queue length is too long. For example, the team may wish to acquire other database server information like: *Average Disk Seconds per*

Figure 9.3 DMAIC from Six Sigma. Source: DMAgIC DMAIC from Six Sigma.

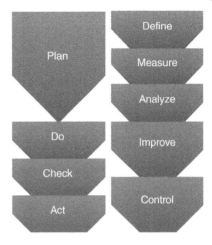

Read, Average Disk Seconds per Write, Percent Disk Time, Average Disk Reads per Second, and *Average Disk Writes per Second*. This would help get to the root cause of the issue. So, essentially this phase is about information collection.

During the "Analyze" phase, the team collects any remaining items required and analyzes the data. The team also creates a baseline with the data for use after the solution is implemented. That way the eventual solution can be evaluated.

Next comes "Improve." Potential solutions are developed. These are evaluated and a best solution selected. Understanding potential side effects and unwanted consequences should be considered during the selection process. Usually an implementation plan and timeline are parts of this phase.

Finally, "Control" takes place. This ensures the solution works and compares new data against the benchmarks to provide evidence. Regular monitoring or long-term measurement may be recommended.

Overall, DMAIC is useful in cloud environments for the same reasons that make it popular in all organizational areas – it is a useful method for organizing problem solving team approaches and offers a structured roadmap from problem definition to solution to long-term control.

TRIZ

The *Theory of Inventive Problem-Solving* or *TRIZ* (the Russian acronym) was developed in the former Soviet Union. Its overall perspective suggests creativity is at the heart of innovation. TRIZ seeks to harness creativity and structure it for use in organizations. The structure makes the process predictable and helps in group situations. TRIZ posits that most problems have been solved. If not, a similar problem has been solved. The key, then, becomes finding the solution and customizing it to the current problem in a creative manner. TRIZ comprises two major concepts: problem generalization and contradiction elimination. Generalization applies to both problems and solutions. TRIZ suggests:

- Both problems and solutions repeat in different industries and technologies.
- Problems are "contradictions" that have been solved before. Therefore, creative solutions can be predicted.

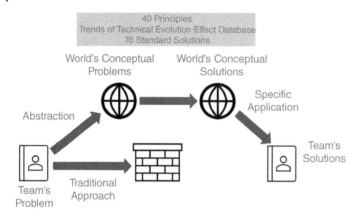

Figure 9.4 TRIZ problem solving methodology. Source: From: Oxford Creativity, TRIZ problem solving methodology.

- Technologies evolve according to repeating patterns.
- Innovation often comes from inspection of patterns from outside the current field.

A team using TRIZ examines repeating patterns to better understand the current contradiction and then finds and customizes the appropriate general solution in a creative way. Figure 9.4 illustrates.

In general, TRIZ starts with the current problem. The team compares it to general TRIZ problems that users have compiled in an open source database with 40 general principles and 76 standard solutions. Often, a matrix is used here. Figure 9.5 provides a portion of this open source resource. Next, the identified solution becomes the starting point for a creative solution. The team solves the problem by eliminating contradictions. Contradictions may be technical (e.g. trade-offs such as speed increases at the expense of power consumption) or physical ones (e.g. a cloud database should be easily accessible but secure against intruders). TRIZ suggests separate principles to help with each contradiction.

Overall, TRIZ is a powerful approach to problem-solving with excellent applications to cloud computing because, as TRIZ suggests, it uses patterns drawing on prior technological innovation in a medium (e.g. the cloud). TRIZ goes far beyond what we have covered in this book but is well worth further exploration.

Microservices

Microservices have gained traction in cloud development approaches. Essentially microservices rethink application architecture in cloud environments. Instead of viewing applications as single units, they are composed of numerous small services (hence the term "micro"). Each service is independent and provides a result or requests information via HTTP (Hypertext Transfer Protocol) or REST application program interface (API). Figure 9.6 illustrates a general representation of the microservice concept.

A microservice approach provides several advantages. First, a variety of languages, approaches, and platforms can be used to develop a microservice since it is an independent

two examples of application:

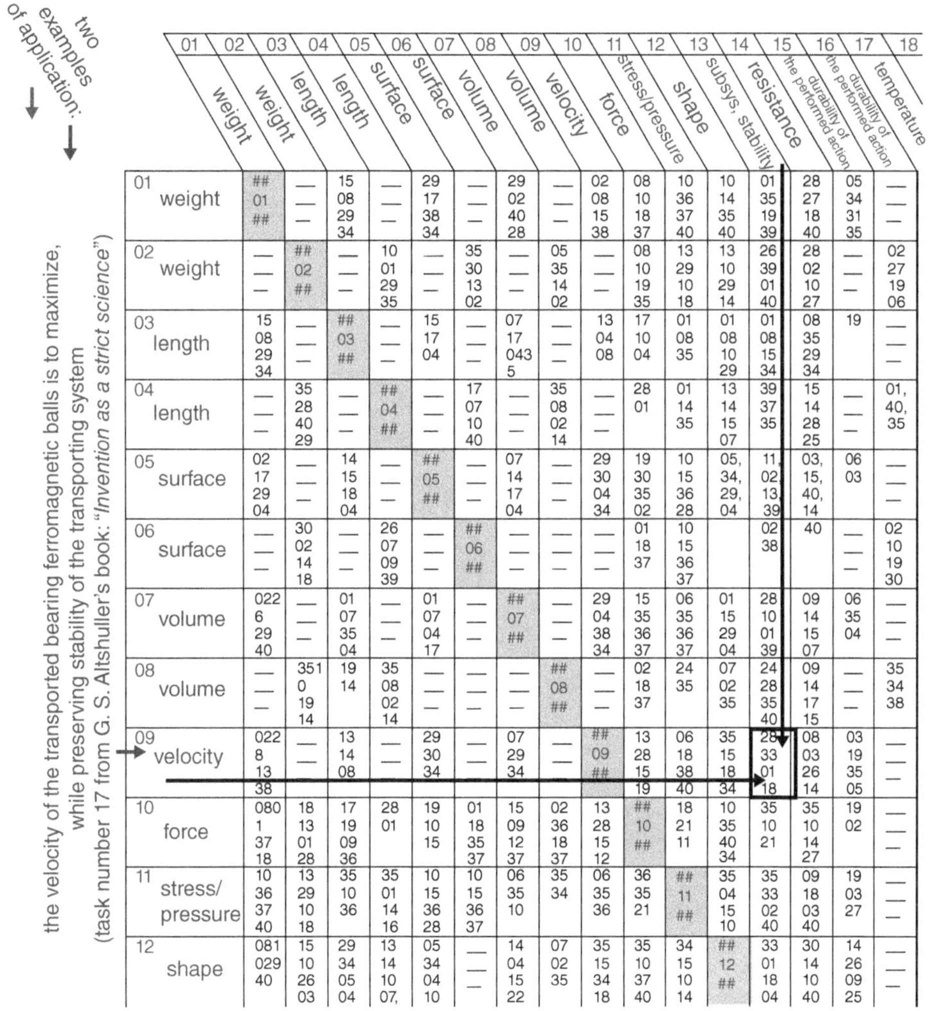

the velocity of the transported bearing ferromagnetic balls is to maximize, while preserving stability of the transporting system

(task number 17 from G. S. Altshuller's book: "Invention as a strict science")

	01 weight	02 weight	03 length	04 length	05 surface	06 surface	07 volume	08 volume	09 velocity	10 force	11 stress/pressure	12 shape	13 subsys. stability	14 resistance	15 the performed action	16 durability of the performed action	17 temperature	18
01 weight	## 01 ##	—	15 08 29 34	—	29 17 38 34	—	29 02 40 28	—	02 08 15 38	08 10 18 37	10 36 37 40	10 14 35 40	01 35 19 39	28 27 18 40	05 34 31 35	—		
02 weight	—	## 02 ##	—	10 01 29 35	—	35 30 13 02	—	05 35 14 02	—	08 10 19 35	13 29 10 18	26 39 01 40	28 02 10 27	—	02 27 19 06			
03 length	15 08 29 34	—	## 03 ##	—	15 17 04	—	07 17 0435	—	13 04 08	17 10 04	01 08 35	01 08 10 29	01 08 15 34	08 35 29 34	19	—		
04 length	—	35 28 40 29	—	## 04 ##	—	17 07 10 40	—	35 08 02 14	—	28 01	01 14 35	14 14 35 15 07	13 37 35	15 14 28 25	—	01, 40, 35		
05 surface	02 17 29 04	—	14 15 18 04	—	## 05 ##	—	07 14 17 04	—	29 30 04 34	19 30 35 02	10 15 36 28	05, 34, 29, 04	11, 02, 13, 39	03, 15, 40, 14	06 03	—		
06 surface	—	30 02 14 18	—	26 07 09 39	—	## 06 ##	—	—	—	01 18 37	10 15 36 37	02 38	40	—	02 10 19 30			
07 volume	02 26 29 40	—	01 07 35 04	—	01 07 04 17	—	## 07 ##	—	29 04 38 34	15 35 36 37	06 35 36 37	01 29 04	28 10 01 39	09 14 15 07	06 04	—		
08 volume	—	35 10 19 14	19 14	35 08 02 14	—	—	—	## 08 ##	—	02 18 37	24 35	07 02 35	24 28 35 40	09 14 17 15	—	35 34 38		
09 velocity	02 28 13 38	—	13 14 08	—	29 30 34	—	07 29 34	—	## 09 ##	13 28 15 19	06 18 38 40	35 15 18 34	26 33 01 18	08 03 26 14	03 19 35 05	—		
10 force	08 01 37 18	18 13 01 28	17 19 09 36	28 01	19 10 15	01 18 35 37	15 09 12 37	02 36 18 37	13 28 15 12	## 10 ##	18 21 11	10 35 40 34	35 10 21	35 10 14 27	19 02	—		
11 stress/ pressure	10 36 37 40	13 29 10 18	35 10 36	35 01 14 16	10 15 36 28	10 15 36 37	06 35 10	35 24	06 35 36	36 35 21	## 11 ##	35 04 15 10	35 33 02 40	09 18 03 40	19 03 27	—		
12 shape	08 10 29 40	15 10 26 03	29 34 05 04	13 14 10 07,	05 34 04 10	—	14 04 15 22	07 02 35	35 15 34 18	35 10 37 40	34 15 10 14	## 12 ##	33 01 18 04	30 14 10 40	14 26 09 25	—		

Figure 9.5 Partial view of TRIZ contradiction matrix. Source: Derived from: FotoSceptyk, Partial view of TRIZ contradiction matrix.

building block of a system. Second, the application's big picture does not need to be shared with developers in mission-critical or sensitive areas. This is especially helpful when software development vendors are used, and an organization wishes to keep their ideas private. Third, new applications can be pieced together quickly, using previously tested microservices. Fourth, operations using microservices are more effective because each service is independently scalable, can be updated without impacting the other services, and can be quickly and individually tested and deployed. Overall, in cloud-based environments using DevOps and agile approaches, this architectural style is an excellent fit.

One the best features of microservices relates to governance. Microservices permit key IT infrastructure elements to be isolated or released with higher levels of security and

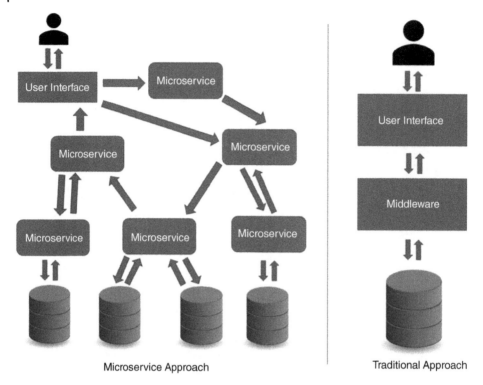

Figure 9.6 Microservices compared to traditional applications.

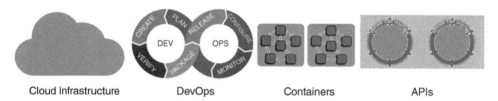

Figure 9.7 Components that help make microservices successful.

additional monitoring capabilities. Data stores can be broken into separate units more easily and subjected to different safety measures. Speed of delivery can be controlled more easily by adding duplicates of microservice components in key geographic regions. This can help solve compliance issues that result from laws in various countries. Scalability becomes granular and only necessary components can be increased in capacity rather than an entire application.

Another feature of cloud computing, containerization, is well suited to microservice usage. As you probably recall, containers enable isolation and easy duplication of virtual IT architecture portions when required. Microservices fit naturally within this framework. Using this approach makes sense from security, economic, and operations standpoints. Figure 9.7 illustrates components that enable a microservices architecture to be successful.

In general, using microservices architecture in the cloud enables DevOps specialists to create faster, scalable, and secure applications. A microservices approach breaks applications into small, independent components that can be built and managed using a variety of tools, techniques, and approaches. The decentralized nature of microservices make good sense in cloud environments.

Cloud Database Applications

Many firms use cloud computing to enhance database applications. In its simplest form, cloud database applications are scalable collections of data, organized, stored, and accessed using software within a cloud environment. Cloud databases take three general forms: traditional databases running in a cloud environment, virtual machine (VM) databases, or database-as-a-service (DBaaS) models. The first form relies on cloud infrastructure managed by an organization's IT staff. The database software is installed and run on cloud infrastructure. This approach is less popular than the other two.

The second database approach uses a VM image of the software provided by the vendor (sometimes the user provides an image). The image may be temporary and used for software testing or other purposes, or it may enhance scalability. If the cloud vendor provides the image, it will probably be optimized for their environment.

The third approach is DBaaS. In this approach, the cloud vendor runs and maintains the database system on their infrastructure and the cloud user obtains accounts and instances of the database for their organization's use. DBaaS vendors manage updates, problems, backups, and security issues. Cloud users pay for use, often on a per access basis.

Cloud databases offer great advantages which mirror general cloud advantages: scalability, reduction of on-premise physical infrastructure, security, high availability, cost savings, and ensuring software and security measures are up to date. For small or medium-sized businesses, DBaaS offers lower entry costs and access to capabilities only available to large corporations in the past. Today's computing world is moving rapidly toward data visualization, Internet of Things, machine learning, and artificial intelligences. DBaaS makes these new technologies accessible to a wider range of businesses.

DBaaS and other cloud databases also present challenges. For instance, data security remains a moving target and requires cooperation between cloud vendor and cloud subscriber. Governance committees must understand the vendor's approach to security and ensure it is sufficient for all compliance and organizational policy needs. Moving data between cloud systems and applications can be tricky and incurs further security concerns. Cloud users may need to use virtual private networks (VPNs) or other approaches to ensure security if their cloud database is used by on-premise application software. Also, database management is a complex task and often requires that an organization has a database specialist on staff.

Cloud Data Models

Cloud vendors generally offer two basic data models: SQL and NoSQL. SQL databases rely on structured tables with rows and columns linked together via relationships composed

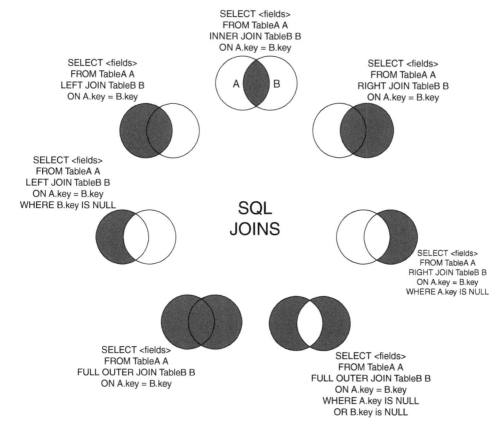

Figure 9.8 SQL joins tables to provide query results. Source: Arbeck, SQL joins tables to provide query results.

of unique, shared key fields. Traditional computing applications use relational database structures and SQL commands to retrieve and manage data. SQL queries retrieve data and provide it to users or application programs. Since relational database systems and SQL were developed prior to cloud environments, these technologies were not specifically intended for distributed use. Figure 9.8 provides an overview of joining tables with SQL commands.

Many database vendors have incorporated features to improve cloud applications, but other approaches emerged. Among these is NoSQL. NoSQL drops the notion of relational tables. Instead, it was developed for cloud-based environments and therefore incorporated features to deal with scalability and heavy read/write demands (e.g. many users obtaining data from a website simultaneously). While this makes sense, NoSQL makes up a small portion of all database applications due to existing developer expertise and because many applications initially were developed prior to NoSQL technologies.

Some relational database vendors have incorporated NoSQL features. This means they added JSON, key-value, graph, and document storage capabilities. Sometimes, these are called *multi-model databases.*

Cloud Database Typical Features

DBaaS provides several features for users. Among these are console screens with tools to provision, configure, and manage database instances, initiate new instances, control users, update, or create database structures, and so forth. Most DBaaS also offer database management tools where the user can perform maintenance on their data and database structure. Tools for managing scalability and costs may also be provided. This helps the user ensure costs, particularly if the database is subject to a "per-access usage" charge, stay under control. Most cloud vendors provide DBaaS in their cloud portfolios. We will review a couple to provide a better idea of this area.

DBaaS Product Examples

Most large vendors provide DBaaS on a subscription basis. The range of specific database packages is broad covering both SQL and NoSQL software. We will look at Amazon's, Microsoft's, and Google's offerings in this area.

Amazon

Amazon is a top database services vendor in the cloud largely due to their success in translating Amazon.com business databases into products offered to others. They provide several database offerings that include relational database (Amazon RDS), NoSQL (Amazon DynamoDB), and a data warehouse (Amazon Redshift). Amazon RDS is considered an industry leader with easy scalability features and high levels of security. It is not meant for a small business with no database experience but rather is meant to be used by an organization with a database administrator.

Microsoft

Microsoft Azure cloud offers several database services including both SQL (SQL Server) and NoSQL (CosmosDB and MongoDB) databases. Like all aspects of Azure, security, and scalability features are strong. Likewise, Microsoft's development tools easily integrate with its database services. Microsoft offers SQL Server Management Studio which gives database users and specialists control over maintaining, changing, updating, and developing databases. Azure is easier for non-specialists to use and often is the choice of smaller businesses as well as large corporations. Business analysts developing solutions find this an excellent choice for flexibility and ease-of-use.

Google

Google provides several DBaaS products and has worked hard to attract business clients. It offers SQL (MySQL) and NoSQL (PostgreSQL) database solutions. It also provides a relational database product called Cloud SQL and a query tool (Big Query) to extract data from its cloud databases. Cloud SQL is considered a good solution for Small and Medium Enterprise (SME)s and functions well as a distributed database system enabling easy access across the web. Google also offers a DBaaS called Cloud Spanner that integrates both SQL and NoSQL features. This tool is aimed at businesses requiring management of complex data types that require high levels of security, replication, and multi-data center support.

Google's DBaaS products are ideal for organizations managing large databases. Like Amazon, Google has experience with vast numbers of requests and transactions. This experience shows in its powerful tools.

Other DBaaS Vendors

Several other vendors offer strong DBaaS products. Many of these services run on the vendor's cloud and are part of a multi-cloud solution used by a client, or can be implemented on a cloud provided by Microsoft, Amazon, or Google.

MongoDB, which is available directly from Azure, also runs on the Atlas MongoDB cloud. This is a hybrid approach because the DBaaS is offered by the same group that built MongoDB. This DBaaS can be implemented on a variety of cloud platforms by Atlas, including Azure, EC2, and Google's Cloud Platform. As mentioned earlier, this is a NoSQL database and offers a wide range of controls, good security, and support from Atlas.

A mainstay of big businesses, IBM offers many cloud database products. Among these, IBM DB2 provides excellent service and a mature product that has been a key element of IT infrastructure of the world's largest firms for decades. DB2 is easy to use and can be operated without a database administrator.

Another powerful database offering comes from SAP. Its Cloud Platform uses an in-memory version of HANA which is specifically developed to power the Internet of Things (IoT) and machine learning applications. Since it is in-memory, it is very fast. It has gained quick acceptance in many large organizations, particularly those that use the SAP ERP system. The SAP Cloud Platform integrates easily with third-party software. It is aimed primarily at large businesses needing to handle huge amounts of data.

Oracle also has a cloud database presence. It offers the Oracle Cloud Platform. Its DBaaS is very secure and offers excellent encryption options. The Oracle DB is known for being easy to use and is integrated with many third-party software packages.

Table 9.2 provides examples of other cloud database products broken down by SQL or NoSQL, and whether they are DBaaS or VM based.

DBaaS is a form of SaaS. That means an organization runs their database application on cloud-based servers. The added complication becomes data storage. The vendor provides software and possesses the organization's data. Therefore, added security and governance issues come into play. A company that uses DBaaS must be sure the vendor offers a product with features that permit:

- Data transfer to another system should it be required.
- Clear backup and recovery procedures.
- Security, encryption, and other safeguards.
- Server provisioning and elasticity based on demand and needs.

Cloud Analytics Services

Data analytics has grown into prominence during the same time frame that cloud computing has become popular. A natural synergy between these areas has emerged and many organizations believe analytics works well using a subscription-based approach. This makes

Table 9.2 Example cloud database products.

Implementation	Virtual machine	DBaaS
SQL	IBM DB2	Amazon RDS
	Ingres	Google cloud SQL
	MySQL	Oracle database cloud service
	Oracle database	Microsoft Azure SQL server
	PostgreSQL	
	SAP HANA	
NoSQL	Apache	Amazon DynamoDB
	Cassandra	Azure DocumentDB
	CouchDB	Google cloud bigtable
	Hadoop	Google cloud datastore
	MongoDB	MongoDB database as a service
		Oracle NoSQL database cloud service

Figure 9.9 Cloud-based analytics receives data from various sources and enables collaboration. Source: Adapted from: Growthlakes, Cloud-based analytics receives data from various sources.

sense for several reasons. First, many analytics projects are single time efforts. This means an elastic service is cost-effective since it is a mainstay of cloud computing. Second, data analytics project results are shared, and a cloud environment can facilitate interaction. Third, data analytics software can be complex to maintain, update, and secure. Figure 9.9 illustrates the idea of cloud-base analytics

Cloud vendors offer these services and remove the burden from the client. Cloud vendors also offer data acquisition tools with APIs and other connections to many sources which saves client time and frustration. For organizations with operational analytics needs, the cloud provides administrative tools, support, and cost reductions. Most cloud analytics platforms offer a variety of features. These features ease administrative overhead and provide resources that reduce client workloads. Among these are:

- State-of-the-art storage resources
- Elastic computing power
- Usage-based cost models
- Data warehouses and other repositories suited to a variety of data types
- Sharing and collaboration tools
- Data model tools
- Analytics tools and statistical tool sets
- Reporting tools and visualization capability
- Governance models and support
- Security features
- Back-up, recovery, and archival tools

Many organizations obtain data from a wide array of sources: on-premise databases, subscription services, web sources, public databases, partners, cloud-based systems, transaction systems, and so forth. This means many cloud users have hybrid needs. The cloud provides an ideal platform for data consolidation. For example, products like Equinix Cloud Exchange enable analytics users to consolidate data from disparate sources into a uniform platform for further processing over a private network.

Most business analytics operations require managing data sources, data models, analysis models, report templates, structures, statistics programs, and reports. Many relevant cloud-based tools are available. We will review several including Microsoft Power BI, Domo, IBM Analytics, and Tableau.

Microsoft Power BI Service

Microsoft's Power BI service is a fully featured business and data analytics tool with capabilities ranging from data acquisition to data modeling to output report generation. It is a cloud-based service that directly interfaces with on-premise desktop versions of Power BI. Power BI has a wide base of users and offers visualization tools from Microsoft and third-party developers. The service automatically extracts data from many sources and offers analysts both exploratory and operational capabilities. Power BI has a wide range of dashboards, interactive online reports, and data management tools.

Power BI is intended to work business environments for data consolidation using standard interfaces. This includes MS-Excel, SQL Server, many database systems, Salesforce, and other applications. Power BI incorporates Excel's more powerful features and Data Analysis Expressions (DAX) scripting to permit advanced data filtering and manipulation. So, it has both drag and drop features and advanced features that work almost like a scripting or programming language.

DAX

DAX is an acronym for Data Analysis Expressions. According to Microsoft, it is a formula language that helps an analyst construct customized expressions to facilitate building "Calculated Columns" and "Measures" in Power BI and other tools. DAX includes many functions like those found in MS-Excel but goes beyond that with specific commands to deal with both relational data and aggregations for many data types. DAX enables the construction of complex statements and filtering logic. Here an example DAX statement from Microsoft:

```
=SWITCH([Month] ,
               1, "January", 2, "February", 3, "March"
             , 4, "April", 5, "May", 6, "June", 7, "July"
             , 8, "August", 9, "September", 10, "October"
             , 11, "November", 12, "December", "Unknown
        month number")
```

Domo

Domo is both a software company and its product used for business intelligence and visualization. It suggests that data holds stories and the tool enables analysts to extract and communicate those stories. Domo features a central dashboard for managing user operations. Users log in to view company data and reports configured either by analysts or themselves depending on their permission levels. The Workbench feature allows data import from many sources and the Analyst feature facilitates report and visualization creation. Domo provides capabilities to generate reports with standard business metrics such as return on investment (ROI) and facilitates display of key performance indicators (KPIs). Domo has an App Store that offers connectors which permit data access from a variety of sources.

IBM Analytics

IBM offers a cloud-based analytics service called *IBM Analytics*. It focuses on business and social media data. It uses artificial intelligence algorithms for predictive analytics. IBM Analytics interfaces to Twitter as a feature of its base package but requires an additional software component to access Facebook and other platforms. IBM Analytics offers many standard visualizations. It also enables users to compile useful information into infographics.

Tableau

Tableau is a very popular visualization software system. It is widely used in sciences, business, economics, and other areas. Tableau offers visualizations that permit users to drill down into the data with increasing levels of detail. Tableau has a cloud-based service called *Tableau Online*. This essentially is a hosted instance of their *Tableau Server* software. This software enables analysts using *Tableau Desktop* to move their dashboards, reports, and visualizations into the cloud where an organization's members can access the data.

Deploying an Analytics Service

The following steps generally need to be followed by organizations wanting to use Analytics as cloud-based services:

- Assessment of user needs.
- IT administrators review architecture and data sources.
- Configure service and software, including security measures and user access levels.
- Data migration including identification of sources. This usually includes data modeling and extraction.
- End-user training.
- Use of tool and development of various dashboards, reports, and visualizations.

Hadoop

The need to process and manage vast amounts of data has pushed many organizations into the use of distributed systems. Tools emerged to address these needs. One of these, Hadoop, is widely used. It is an open source data processing and storage framework for distributed systems. Hadoop often provides the central data processing infrastructure for organizations engaged in data mining, business analytics, predictive analytics, machine learning, and other techniques that involve big data applications. Hadoop was developed as it became apparent that datasets in distributed environments could become much larger than current processing systems' capabilities. Since the data would be larger than any one physical device, the logical solution became linking smaller, less expensive devices.

Hadoop comprises four primary modules. These are *HDFS* (Hadoop Distributed File System), *MapReduce*, *Hadoop Common*, and *YARN* (Yet another Resource Navigator). Each module performs a unique big data task in the overall framework. HDFS supports high speed access for application data. Hadoop's objective is to ensure that large amounts of data can economically be stored across low-cost distributed computing systems. HDFS stores the data in a manageable form on linked storage devices. MapReduce performs two basic tasks. One is pulling data from the database and the second is mapping the data into a form that suits the current analysis which could mean performing an operation on the data (e.g. counting or averaging values). So, putting into a form suitable for the analysis is the *map* part and performing an operation *reduces* the data. Hadoop Common provides a Java-based toolset which analysts can use for reading data or other tasks, suited to the end-user's computer system. Finally, YARN manages and controls the storage resources available to Hadoop.

Hadoop in the Cloud

Technically Hadoop can run in the cloud, but it was developed prior to the advent of cloud computing. Both cloud and Hadoop architectures are distributed but Hadoop was meant for on-premise or at least physical data centers which cloud environments virtualize to permit elasticity and VM provisioning on the fly.

Hadoop has a weakness that makes it less desirable as a cloud-based application: security. It was not intended to run in cloud environments, so its security measures are not adequate to safeguard against data breaches. It was developed to permit an array of off-the-shelf computers (or in some cases a series of older computers that an organization already possessed) to be used for cheap data storage.

However, many organizations use cloud-based implementations of Hadoop either in a pure sense or as a hybrid model in conjunction with their physical data centers. In general, Hadoop can work in the cloud, either as a service, vendor offered, or self-deployed.

Several vendors offer key components of Hadoop as services. For example, Amazon provides Elastic MapReduce and Azure's HDInsight offers MapReduce within its secure infrastructure. This means Hadoop users can utilize its functionality without installing Hadoop in their cloud environment. Another vendor-based approach is to offer pre-installed, complete Hadoop instances in their cloud environments. For instance, Cloudera CDH is available on AWS, Azure, Rackspace, and other public clouds. This gives the user an entire implementation of Hadoop in the cloud with security and elasticity already in place. Finally, organizations have implemented Hadoop on their IaaS public or private clouds. This means IT specialists must manage their own Hadoop clusters. Some organizations have adapted this to a hybrid environment where some storage is on-premise, and some is cloud-based.

In all three cases, having Hadoop's capabilities can be vital depending on organizational needs. However, remember that Hadoop provides a powerful solution for big data management, but it was developed in pre-cloud environments. Hadoop also lacks in several other areas. For example, it does not support SQL-based analysis well nor does it offer easy indexing options. It is not a good solution to real time data analysis. For an organization deciding its future direction regarding big data management, other cloud-based options exist that were not available when many organizations moved to Hadoop. Among these are Hadoop's sister products, Spark and Storm; and, Google's BigQuery.

Apache Spark

Spark is in wide use as a computational engine in Hadoop environments. Even though it was developed by Apache (who developed Hadoop originally), it does not need Hadoop to operate. This makes it ideal for cloud implementation where it can be used within other data environments. In fact, many organizations and cloud vendors have structured their analytics tools around Spark. Like Hadoop, Spark is open source and processes data up to 100 times faster than MapReduce. In emerging machine learning environments, Spark is an excellent choice. Figure 9.10 looks at Spark's components.

Apache Storm

Storm solves problems with Hadoop. Like Spark, it was developed and is maintained by Apache and its user communities. Storm is a real time data processing tool, as opposed to Hadoop's batch processing approach. Storm was engineered to operate in environments where data streams have no discrete ending point. This is ideally suited for IoT or situations where data may sporadically appear in rapid spates.

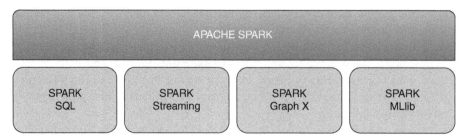

Figure 9.10 Components of Apache Spark. Source: Modified from Apache, Components of Apache Spark.

Figure 9.11 Logo of the OpenStack® project as of 2016. Source: http://www.openstack.org/brand/openstack-logo/logo-download Author: OpenStack Community. For educational use only. Also available on Wikipedia at https://en.wikipedia.org/wiki/OpenStack

---

Google BigQuery

Google's BigQuery platform is a complete data analytics environment intended for use in big data applications. It employs SQL to Google storage in the cloud and provides analysts with numerous options regarding ways to retrieve and report data. It offers numerous data mining tools for uncovering patterns in online data sets.

Open Source Private Cloud Software

Much of our book has focused on cloud computing services offered by major vendors such as Microsoft, Amazon, and Google. Early in the book, we talked about private clouds and that cloud technology can be implemented by an organization using its own data center. This requires an experienced and full-time IT staff ready to take on all the security, governance, and technical challenges that cloud vendors usually manage. Vendors that provide private cloud software are: VMWare, Microsoft Cloud, SAP Cloud Platform, Dell Technologies Cloud, Amazon Virtual Private Cloud, and Google Virtual Private Cloud. There is one rather unique option that we will focus on – OpenStack.

OpenStack

OpenStack is a popular, open-source cloud software platform (sometimes called a cloud operating system). Figure 9.11 is OpenStack's logo. Like all the cloud systems we have covered, OpenStack permits IT specialists to control storage, compute, networking, and other cloud resources through a central dashboard. In general, OpenStack provides the tools needed for an organization to build and deploy a cloud infrastructure.

Many IT experts believe OpenStack is ideal for DevOps environments because it puts the deployment and control of the software (and its underlying source code) in the hands of the team managing its use within an organization. This means it can be customized to

Table 9.3 Sample of OpenStack's primary components.

Component	Purpose	Use
Cinder	Block storage	Traditional file and object storage that places items in specified locations for access speed requirements.
Glance	Virtual image services	Permits templating of virtual images and making them accessible to future deployments.
Heat	Orchestration	Permits creation of a file with all needed components listed for deployment of instances of resources. Enables a cloud application to be quickly created in identical ways each time.
Horizon	Dashboard	GUI permitting access to all components within OpenStack plus tools to manage the entire cloud.
Keystone	Identity services	Allows users and resources to be mapped and permissions granted.
Neutron	Networking	Ensures communication among all components in cloud system.
Nova	Compute engine	Deploys VMs to manage processing needs and elasticity. Manages large numbers of virtual instances of resources.
Swift	Storage system	Determines where to store files and other objects across all resources. Deals with storage scalability.

Source: Adapted from OpenStack, Sample of OpenStack's primary components.

meet the needs of an organization using an agile approach. Its source code is managed by the OpenStack community. OpenStack offers many platform access and management tools. It provides a Graphic User Interface (GUI) dashboard, a command line interface, and many APIs.

OpenStack Components

Since OpenStack is open source software, IT specialists can start with the code repository to choose what their organization needs. Any missing components can be added as custom software modules by developers. Table 9.3 and Figure 9.12 illustrate.

OpenStack and its software are developed and managed by a user group and overseen by the OpenStack Foundation. This non-profit organization works to keep an active community of developers organized and motivated to work on the project and incorporate enhancements.

Other Services

We have described several services that run in the cloud, but others exist in a wide range of categories. In fact, an incredible number of vendors currently work to enhance many services in cloud environments. These range from compute services to applications such as email. We will briefly look at several sample services, but many more exist, particularly in SaaS.

Figure 9.12 OpenStack comprises several main services and components. Source: OpenStack Foundation.

Compute Services

We have mentioned *compute services* indirectly throughout this book but want to take a moment to specifically focus on this term because it is widely used, especially by Amazon. Compute services focus on applications requiring central processing unit (CPU) intensive operations as opposed to data intensive operations. Compute tasks in the cloud are driven by virtual CPUs, arithmetic processing units (APUs), and graphical processing units (GPUs). Rendering 3-D graphics is a particularly intensive task on a computer and often requires powerful compute resources. As you might imagine, online games, and virtual 3-D environments fall into this category.

In general, compute services are those that require a great deal of computational power and therefore require resources adequate to accomplish those tasks. A utility computing approach which treats the cloud as a pay-as-you-go elastic model depends heavily on compute resources and spins up VMs whenever demand requires.

Application Services

Application services are one of the fastest growing areas in cloud computing. These services often fall under the category of SaaS and include application programs used to accomplish anything that an end-user might wish to do. In general, cloud application services run from servers located in a cloud-based data center. Most often, application services are off-premise and are shared resources used by large numbers of end-users which are permitted use according to their authorization level. Cloud applications services include email, calendaring, business software, accounting systems, tax software, and so forth. The list is nearly endless. Cloud application services are different from pure web applications primarily regarding their underlying architecture, although many cloud providers use web technologies to offer their services. For a simple example, think of Office 365. It runs on a desktop computer and uses cloud-based resources for retrieving and storing files, updates, and other tasks. Gmail, a cloud-based email system, on the other hand uses a web-based interface to provide access to email messages.

Summary

Chapter 9 provided a look at how IT activities and approaches are impacted by cloud computing. The very concept of DevOps, while not completely dependent on the cloud, is enabled and promoted by cloud computing. IT specialists, more than ever, integrate development and operations actions and monitor the immediate impact on resources, users, and security functions.

We also reviewed how many IT resources are enhanced in cloud environments. This included database services, data analytics, and big data operations. Depending on the needs of an organization, open source software platforms may be adapted to specific business models. This could mean using a private cloud operating system or putting together an infrastructure from several open source pieces. Overall, the power of cloud computing and its transformational nature is very apparent.

References

Crosby, P.B. (1979). *Quality Is Free: The Art of Making Quality Certain*. New York: McGraw-hill.

Lewis, W.E. (2017). *Software Testing and Continuous Quality Improvement*. CRC press.

Swaminathan, B. and Jain, K. (2012). Implementing the lean concepts of continuous improvement and flow on an agile software development project: An industrial case study. In: *2012 Agile India*, vol. 2012, 10–19. IEEE.

Walton, M. (1988). *The Deming Management Method: The Bestselling Classic for Quality Management!* Penguin.

Warkentin, M., Sayeed, L., and Hightower, R. (1997). Virtual teams versus face-to-face teams: An exploratory study of a web-based conference system. *Decision Sciences* 28 (4): 975–996. https://doi.org/10.1111/j.1540-5915.1997.tb01338.x.

Further Reading

April, A. and Abran, A. (2012). *Software Maintenance Management: Evaluation and Continuous Improvement*. Wiley.

Bass, L., Weber, I., and Zhu, L. (2015). *DevOps: A Software architect's Perspective*. Addison-Wesley Professional.

Lehner, W. and Sattler, K.U. (2010). Database as a service (DBaaS). In: *2010 IEEE 26th International Conference on Data Engineering (ICDE 2010)*, 1216–1217. IEEE.

Lwakatare, L.E., Kuvaja, P., and Oivo, M. (2015). Dimensions of DevOps. In: *International Conference on Agile Software Development*, 212–217. Cham: Springer.

Parker, Z., Poe, S., and Vrbsky, S.V. (2013). Comparing NoSQL MongoDB to an SQL DB. In: *Proceedings of the 51st ACM Southeast Conference*, 1–6. ACM.

Sahri, S., Moussa, R., Long, D.D., and Benbernou, S. (2014). DBaaS-expert: A recommender for the selection of the right cloud database. In: *International Symposium on Methodologies for Intelligent Systems*, 315–324. Cham: Springer.

Stonebraker, M. (2010). SQL databases v. NoSQL databases. *Communications of the ACM* 53 (4): 10–11.

Xie, J. and Li, F. (2009). Study on innovative method based on integrated of TRIZ and DMAIC. In: *2009 International Conference on Information Management, Innovation Management and Industrial Engineering*, vol. 1, 351–354. IEEE.

Zulkernine, F., Martin, P., Zou, Y. et al. (2013). Towards cloud-based analytics-as-a-service (CLAaaS) for big data analytics in the cloud. In: *2013 IEEE International Congress on Big Data*, 62–69. IEEE.

Website Resources

Data Analytics Tools

ClouderaCDH: https://www.cloudera.com/products/open-source/apache-hadoop/key-cdh-components.html

Domo: http://www.domo.com

EquinixCloudExchange: https://www.equinix.com/interconnection-services/equinix-fabric/

Hadoop: https://hadoop.apache.org

IBMAnalytics: https://www.ibm.com/analytics

MicrosoftPowerBI: https://powerbi.microsoft.com/en-us

Tableau: https://tableau.com

DBaaS

NoSQL

AmazonDynamoDB: https://aws.amazon.com/dynamodb

AzureCosmosDB: https://azure.microsoft.com/en-us/services/cosmos-db

GoogleCloudBigtable: https://cloud.google.com/bigtable

GoogleCloudDatastore: https://cloud.google.com/datastore

MongoDBDatabaseasaService: https://www.mongodb.com/cloud/atlas

OracleNoSQLDatabaseCloudService: https://www.oracle.com/database/nosql-cloud.html

SQL

AmazonRDS: https://aws.amazon.com/rds/

GoogleCloudSQL: https://cloud.google.com/sql

OracleDatabaseCloudService: https://www.oracle.com/in/database/cloud-services.html

MicrosoftAzureSQLServer: https://azure.microsoft.com/en-us/services/sql-database

DevOps

Chef: https://chef.io
GitHub: https://github.com
MicrosoftVisual Studio: https://visualstudio.microsoft.com
Nagios: https://www.nagios.org
Puppet: https://puppet.com
SharePoint: https://www.microsoft.com/en-us/microsoft-365/sharepoint/collaboration
Trello: https://trello.com
Zoom.us: https://zoom.us

Hadoop Competitors

GoogleBigQuery: https://cloud.google.com/bigquery
Spark: https://spark.apache.org
Storm: https://storm.apache.org

Private Clouds

AmazonVirtualPrivateCloud: https://aws.amazon.com/vpc
DellTechnologiesCloud: https://www.delltechnologies.com/en-us/cloud/index.htm
GooglePrivate Cloud: https://cloud.google.com/vpc
MicrosoftCloud: https://azure.microsoft.com/en-us/free/hybrid-cloud
OpenStack: https://www.openstack.org
SAPCloudPlatform: https://www.sap.com/products/cloud-platform.html
VMWare: https://www.vmware.com

Virtual Databases

NoSQL

ApacheCassandra: https://cassandra.apache.org
CouchDB: https://couchdb.apache.org
Hadoop: https://hadoop.apache.org
MongoDB: https://www.mongodb.com

SQL

IBM DB2: https://www.ibm.com/products/db2-database
Ingres: https://www.actian.com/data-management/actian-x-hybrid-rdbms
MySQL: https://www.mysql.com
Oracle Database: https://www.oracle.com/database
PostgreSQL: https://www.postgresql.org
SAP HANA: https://www.sap.com/products/hana.html

10

What Is the Cloud Future?

Without a doubt, the future of business computing (and personal computing) resides in cloud-based environments. The cloud is a relatively new technology and its full potential remains to be reached. Most businesses recognize the value in cloud computing and view it as a solution to many of their challenges–technical, managerial, and financial.

Aside from "more of the same" advantages the future of cloud computing offers, such as elasticity, monetary savings, reduction of on-premise data center expenses, and so forth, many exciting developments and new ideas emerge with cloud computing. Among these are NoOps, Zero Knowledge Clouds, Serverless Architectures, Machine Learning, Cloud Streaming Services, Edge computing, and others. We will look at several cloud-based trends emerging to help business leaders and cloud subscribers envision their future.

NoOps

NoOps is an abbreviation for *No Operations*. This idea suggests IT infrastructures in organizations could become so automated that operations personal would no longer be needed. As you might imagine, this idea sends chills up the spine of people with careers in IT operations and has caused more than a few Twitter storms over the past year.

The NoOps concept recognizes someone must set up the automation but once running, a dedicated IT operations staff is not needed. A more realistic way to describe this trend would be as: *more automation*.

Without a doubt, the cloud's future includes more automation, and intelligent software systems that abstract organizational needs from usage patterns and translate that into operational environments. Automation is useful for anomaly detection and understanding when resources are compromised or used for unintended purposes. In the near term, with the DevOps movement, and powerful automation and orchestration, fewer operations personnel will be required. Instead, new roles related to integrating secure cloud operations with business needs will emerge. In the more distant future, perhaps machine learning, AI, and other technologies will provide most IT operations services. However, organizational needs, SLAs, and other business-side processes will remain within the IT operations specialist domain for a long time to come.

Cloud Technologies: An Overview of Cloud Computing Technologies for Managers, First Edition. Roger McHaney.
© 2021 John Wiley & Sons Ltd. Published 2021 by John Wiley & Sons Ltd.
Companion website: www.wiley.com/go/mchaney/cloudtechnologies

Everything as a Service (EaaS)

As mentioned in Chapter 9, cloud-based application services have grown at an incredible rate. Practically everything in computing environments now runs in the cloud. This *EaaS* trend is not slowing. Most new activity relates to business application software and consumer products such as video and music streaming, and mobile apps.

Services that rely on subscription charges have been embraced by business and consumer alike. Netflix and other vendors demonstrate that steady cash flows appeal to investors. From a business perspective, costs are predictable and managed more easily in these environments which makes cloud-based services even more attractive.

Zero Knowledge Cloud Storage

Security is important regarding the cloud's future. With regular news about data breaches and cyber-criminals stealing millions of consumer records, this area must be considered for future enhancements.

Zero-knowledge cloud storage is an idea that makes cloud computing more secure. Although this has been available for some time, it suffers from speed and management issues which prevent some organizations from using this ultra-secure method of storing data. For many companies, only a small portion of their most sensitive data is stored this way.

So, what is zero-knowledge cloud storage? It simply means all data is encrypted prior to movement into the cloud so the cloud provider and anyone working in their data center cannot access the data in a meaningful or useful way. While it offers higher security, several challenges exist, and several future solutions will emerge to make zero-knowledge cloud storage a much wider practice. First, encryption keys in zero-knowledge storage are kept by the data owner only. Therefore, if the key is lost or corrupted, data cannot be retrieved in a useful way. If the key is lost, the data is gone for all practical purposes.

The other downside relates to storage and transmission speeds. Since encryption and decryption operations take place prior to cloud storage, operations are slowed. This makes real time use of this technology unviable in many applications. Cloud vendors hope to speed up zero-knowledge cloud storage as well as find better approaches to accomplish the same outcome–highly secure data storage. Without a doubt many new developments will emerge in this area.

Serverless Architecture

Serverless architecture is a computing model receiving renewed attention among large cloud providers (e.g. Google, Amazon, and Microsoft), and third-party vendors hoping to develop a new market. Serverless technology is consistent with the idea of NoOps, where cloud clients do not worry about day-to-day operations in their data centers. Google already uses similar technology to power its largest applications like Gmail and YouTube. Serverless architecture takes the concept of spinning up new servers when demand requires (or deprovisioning underutilized servers) and automates it. So, serverless architecture refers to the concept of dynamically automating the level of infrastructure needed and then provisioning (or deprovisioning) the resources required in real-time to keep operations running optimally.

The concept does have challenges such as latency (e.g. the delay between resource need and resource availability), security, and privacy. Special care must ensure remnants of deprovisioned servers and critical data are not languishing in cloud memory. However, with DevOps gaining popularity, using serverless architecture appears to be gaining momentum.

Multicloud

Large organizations often rely on business models using resources from multiple vendors. For instance, the best-in-class application software may come from one source (e.g. Sales-force or SAP), productivity software from another (Office 365), and so forth. This means an organization may already deal with multiple public clouds in its IT infrastructures. In addition, many organizations hesitate to rely on a single cloud provider for all needs and do not wish to put all "eggs in a single basket." *Multicloud* is not the same as a hybrid cloud which relies on using a combination of on-premise, private, and public clouds. Instead, multicloud engages multiple cloud vendors. The motivation for going with a multicloud approach is complex. It can reduce cloud subscriber reliance on one vendor to provide better leverage when negotiating future SLAs. It could alleviate governance concerns and ensure business needs are met. It might increase computing or communication speed due to geographic proximity or because one mission critical application requires higher speed and it is too expensive to deploy the entire organizational computing platform that way. In short, the idea of picking and choosing clouds from a variety of sources can add business value and will undoubtably continue into the foreseeable future.

Small Business Clouds

Cloud computing continues to grow in importance for small and medium enterprises (SMEs). This is true particularly for start-ups and new cash-strapped innovations seeking to minimize capital expenditures. In fact, many new cloud computing models such as NoOps and subscription-based services ensure costs are based on consumption and computing services easily scale up as an enterprise grows. Many SMEs lack the ability to hire and maintain an IT staff and using cloud computing with data center services can replace this need in the short term, and in some cases even for the long term. Top-notch help desk and IT specialists are available through cloud vendors. Also, a SME can extend its reach into global markets using cloud-based IT services.

Machine Learning

Machine learning (ML) is an exciting area in both cloud computing and the broader IT world. In general terms, ML involves development of software that becomes more accurate in predicting outcomes without the intervention of humans. In other words, algorithms in ML find better ways of achieving goals and improve over time.

Several different approaches are used in ML including supervised and unsupervised learning. Supervised learning includes a more significant human dimension in the process where data scientists provide feedback to the model as it develops and determine what variables should influence the learning behavior. Supervised learning involves a training phase prior to use on a full data set. See Figure 10.1.

Unsupervised learning, on the other hand, uses algorithms and iterative cycles to explore data and arrive at recommendations. Often, unsupervised learning involves neural networks and solves complex problems in domains such as speech recognition, image classification, and those that involve rich media.

ML provides exciting possibilities for cloud computing and fits nicely into the mix with DevOps, NoOps, and serverless computing architectures. For example, in the IaaS arena, longitudinal data can determine the best time to provision a new VM, so no latency issues are experienced by users. Likewise, patterns of resource use can inform an ML algorithm whether a security breach is likely or not. On the SaaS side, applications can use ML to look at email patterns to determine if a customer is likely to leave for a competitor, or ML can help determine the best time to purchase goods and services in a manufacturing operation. ML also can be useful in powering help desk applications. The uses of ML are nearly endless as this technology matures and moves into the mainstream.

Major cloud vendors all offer ML packages for inexpensive deployment. These packages interface with cloud-based data collection and provide insights to help identify issues, predict needs, or discover ways of economizing. The packages generally are deployed as SDKs and APIs for use with application software and database systems. Using ML currently requires enterprise expertise and may necessitate employing a data scientist or business analyst.

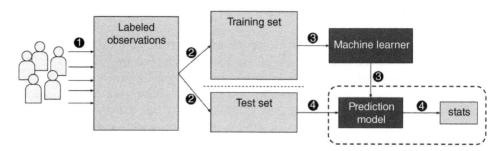

Figure 10.1 Supervised machine learning process. Source: EpochFail, Supervised machine learning process.

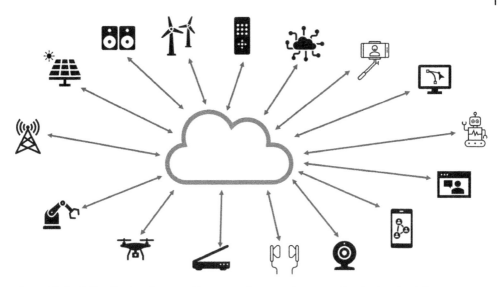

Figure 10.2 IoT pulls data from a wide range of sources for processing in the cloud. Source: Pixabay, IoT pulls data from a wide range.

Internet of Things (IoT)

Cloud computing is a natural fit with the *Internet of Things (IoT)*. Large data sets are pulled into the cloud from a wide variety of devices. This is compatible with cloud attributes such as scalable storage, variable processing power, and analysis tools. Devices ranging from personal fitness bands to home assistants like Amazon's Echo to GPS devices tracking automobile movement provide rich data sources for organizations seeking to leverage ML and other techniques to improve operations or more effectively interact with customers and business processes. Additional data comes from social media, sensors, and mobile devices. The cloud provides required computing power and storage capacities while IOT offers new ways to pull data from everyday devices in ways that seem like something out of science fiction. This technology provides a glimpse of its future business potential and plays a key role in organizational planning. Figure 10.2 illustrates.

Cloud Computing as a Utility

Some experts foresee cloud computing moving to a greater utility-type model as time goes on–both for private individuals and organizations. In many ways, cloud computing already operates as a utility, but will become more so with automatic provisioning and service option choices. The model is like electric utilities or city water metering. As computing resources are consumed, a monthly fee is assessed and paid by the user. This approach promises that shared resources reduce costs and improve power for everyone in the long run.

Cloud Streaming Services

Cloud streaming services are set for big changes and functional improvements. Consumer demand for streaming services has increased rapidly and expectations for entertainment videos, music, television channels, virtual reality applications, video games, online courses, digital books, and other material continues to increase. Businesses require content servers for training material, organizational resources, customer help material, and digitalized records. Individuals want to play games, watch movies, listen to music, read books, and interact with peers using cloud-based systems. These areas are certain to continue growing and placing demands on network bandwidth and storage resources.

Edge Computing

Edge computing promises to add a layer of computing power that enhances the nexus between cloud computing and IoT. In its most basic form, Edge computing relies on technologies that allow computation to occur on a network's edge–usually on downstream cloud data and upstream IoT data (Shi et al. 2016). See Figure 10.3. Basically, essential data processing occurs closer to IoT devices or other sources via a mobile, Bluetooth, or wireless

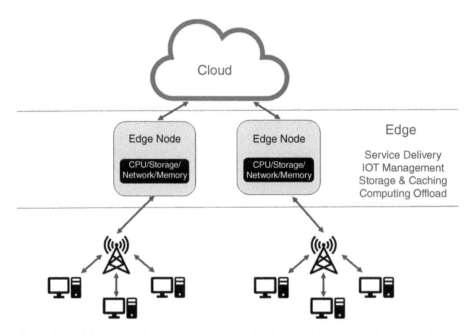

Figure 10.3 Edge computing architecture example. Source: NoMore201, Edge computing architecture example.

network. According to De Donno et al. (2019), Edge computing is implemented in various ways. Among these are *Mobile Cloud Computing (MCC)* and *Cloudlet Computing (CC),* or *Mobile Edge Computing (MEC).* At its core, MCC advocates mobile devices offload storage and computation to reduce the processing load. This idea suggests the best candidates for processing and storage are devices at a network's edge. This enables pre-processing prior to data moving into a broader cloud environment. CC is a leading implementation of MCC to enable IoT's integration with the cloud. According to De Donno et al. (2019), "[a] cloudlet is a trusted, small Cloud infrastructure, located at the edge of the network and available for nearby mobile devices that collaborates with the Cloud to compute the results and then sends them back to mobile devices." Another approach to Edge computing is MEC. This brings computation and storage to a mobile network's edge via a Radio Access Network. This offers context awareness, scalability, location responsiveness, low latency, and high bandwidth potential.

Fog Computing

Fog computing takes Edge to a higher level. It offers distributed computing, storage, control, and networking capabilities closer to the user (Chiang et al. 2017). According to De Donno et al. (2019) this is:

> the highest evolution of the Edge computing principles. Indeed, Fog computing is not limited to only the edge of the network, but it incorporates the Edge computing concept, providing a structured intermediate layer that fully bridges the gap between IoT and Cloud computing. In fact, Fog nodes can be located anywhere between end devices and the Cloud, thus, they are not always directly connected to end devices. Moreover, Fog computing does not only focus on the "things" side, but it also provides its services to the Cloud. In this vision, Fog computing is not only an extension of the Cloud to the edge of the network, nor a replacement for the Cloud itself, rather a new entity working between Cloud and IoT to fully support and improve their interaction, integrating IoT, Edge and Cloud computing.
>
> (p. 150936)

Fog computing utilizes three-tier architecture as shown in Figure 10.4. The IoT Layer represents geographically dispersed devices, smart appliances, smart homes, drones, industrial sensors, crop moisture sensors, or any other IoT device.This interacts with the Fog layer where Fog nodes compute, communicate, and store data temporarily. Finally, the Cloud layer receives processed data into its centralized infrastructure where PaaS or IaaS processes integrate, store, and use the data. Fog provides several benefits to the interface between IoT and Cloud Computing. These advantages have been called SCALE: Security, Cognition, Agility, Latency, Efficiency and represent how preprocessing closer to the user or device improves cloud operations (Chiang et al. 2017; De Donno et al. 2019).

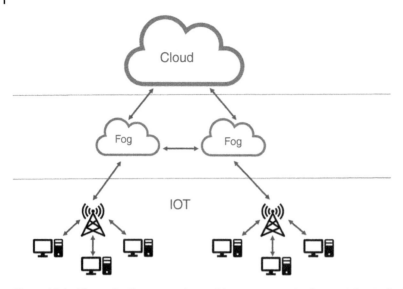

Figure 10.4 Three-tier Fog computing architecture example. Source: Michele De Donno, Three-tier Fog computing architecture example.

Summary

Commercial cloud computing has existed for about a decade and has dramatically altered organizational approaches to IT infrastructure and application delivery. Cloud computing has triggered enormous innovation, new products, and ideas. The very thought that data centers could become virtual entities located thousands of miles from user bases is incredible, yet it has happened everywhere.

The next generation of cloud computing shows no signs of slowing down or stagnating. Technologies like machine learning, serverless architecture, AI, and IoT promise advances even more incredible than what we have today. The convergence of organizational and private computing demands has triggered innovations in streaming technologies, virtualization, and Internet speeds. New ways to manage and control costs produce innovative SLAs between companies and cloud providers that offer stability, security, and frameworks for sound governance. Management of IT has integrated development and operations activities with DevOps and NoOps philosophies. This breaks down barriers and offers new ways to enhance organizational effectiveness.

Additional functions are virtualized and moved into elastic environments responding to needs rapidly and in ways that inhibit waste. Advanced monitoring systems collect data and interact with ML algorithms to prevent problems, security breaches, and operating faults. Edge computing moves processing closer to the source making IoT faster and better. Business analytics provide reports to decision makers and increasingly are enhanced with AI-based recommendations.

All in all, the cloud's future is bright and will impact EVERYONE but especially those in IT related fields (Figure 10.5).

Figure 10.5 Future of cloud computing is bright!. Source: Public domain files. Future of cloud computing is bright. http://www.publicdomainfiles.com/show_file.php?id=13534699217949.

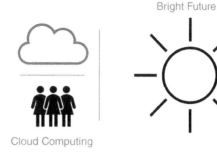

Bright Future

Cloud Computing

References

Chiang, M., Ha, S., Risso, F. et al. (2017). Clarifying Fog computing and networking: 10 questions and answers. *IEEE Communications Magazine* 55 (4): 18–20.

De Donno, M., Tange, K., and Dragoni, N. (2019). Foundations and evolution of modern computing paradigms: cloud, IoT, Edge, and Fog. *IEEE Access* 7: 150936–150948.

Shi, W., Cao, J., Zhang, Q. et al. (2016). Edge computing: vision and challenges. *IEEE Internet of Things Journal* 3 (5): 637–646.

Further Reading

Chowdhury, A., Karmakar, G., and Kamruzzaman, J. (2019). The co-evolution of cloud and IoT applications: recent and future trends. In: *Handbook of Research on the IoT, Cloud Computing, and Wireless Network Optimization*, 213–234. IGI Global.

Cicconetti, C., Conti, M., Passarella, A., and Sabella, D. (2020). Toward distributed computing environments with serverless solutions in edge systems. *IEEE Communications Magazine* 58 (3): 40–46.

Gusev, M. and Dustdar, S. (2018). Going back to the roots – the evolution of Edge computing, an IoT perspective. *IEEE Internet Computing* 22 (2): 5–15.

Lu, Y. (2019). Artificial intelligence: a survey on evolution, models, applications and future trends. *Journal of Management Analytics* 6 (1): 1–29.

Moreno-Vozmediano, R., Montero, R.S., Huedo, E., and Llorente, I.M. (2019). Efficient resource provisioning for elastic cloud services based on machine learning techniques. *Journal of Cloud Computing* 8 (1): 5.

Parra, G.D.L.T., Rad, P., Choo, K.K.R., and Beebe, N. (2020). Detecting internet of things attacks using distributed deep learning. *Journal of Network and Computer Applications*: 102662.

Rad, P., Roopaei, M., Beebe, N., Shadaram, M. and Au, Y. (2018, January). AI thinking for cloud education platform with personalized learning. *Proceedings of the 51st Hawaii international conference on system sciences*.

Roman, R., Lopez, J., and Mambo, M. (2018). Mobile edge computing, Fog et al.:a survey and analysis of security threats and challenges. *Future Generation Computer Systems* 78: 680–698.

Satyanarayanan, M., Bahl, P., Caceres, R., and Davies, N. (2009). The case for VM-based cloudlets in mobile computing. *IEEE Pervasive Computing* 8 (4): 14–23.

Sehgal, N.K., Bhatt, P.C.P., and Acken, J.M. (2020). Future trends in cloud computing. In: *Cloud Computing with Security*, 235–259. Cham: Springer.

Yao, Y. (2019). Emerging cloud computing services: A brief opinion article. In: *Cloud Security: Concepts, Methodologies, Tools, and Applications*, 2213–2218. IGI Global.

Glossary

Chapter 1 List of Terms

Capacity Planning The process used to determine and fulfill future demands expected for an organization's IT resources, duties, operations, and services.

Client-Server Approach Computing power exists both at the user's terminal and on servers where data and applications respond to requests from individual users.

Cloud Computing Delivering computing services – which may include servers, databases, storage, networking, software, data analytics, security solutions, organizational systems, virtual computers, and much more – over the Internet.

Elasticity Feature of cloud computing where resource provisioning responds to business changes by upsizing or downsizing capabilities to match the current demand.

Enterprise Cloud A cloud hosted on hardware systems located within an organization's data center. A private cloud is essentially an intranet where all the data and services are protected by a firewall. Also called a private or internal cloud.

Granularity Describes the service level of software components. High granularity indicates breaking components down to a finer level.

Hybrid Cloud Combines features of both public and private clouds, using technology that permits sharing data and applications between them.

IaaS Approach The lowest level cloud solution. It focuses on system configuration issues. Specifically, IaaS clouds host infrastructure components traditionally managed in on-premise data centers. IaaS offers services that include servers, data storage, networking hardware, and virtual machines. IaaS seeks to be a fully outsourced service that replaces a data center and offers pre-configured hardware (or software) through an interface. Customers install software and services on the IaaS cloud and run/manage their applications as if it were an on-premise data center.

Infrastructure as a Service (IaaS) Hardware is supplied and managed by an external party.

Internal Cloud A cloud hosted on hardware systems located within an organization's data center. An internal cloud is essentially an intranet where all the data and services are protected by a firewall. Also called a private or enterprise cloud.

Cloud Technologies: An Overview of Cloud Computing Technologies for Managers, First Edition. Roger McHaney.
© 2021 John Wiley & Sons Ltd. Published 2021 by John Wiley & Sons Ltd.
Companion website: www.wiley.com/go/mchaney/cloudtechnologies

Lag Strategy Capacity planning approach that adds IT resources only after capacity is reached. This approach can suffer from under-provisioning and being unable to rapidly react to organizational needs.

Lead Strategy Capacity planning approach where IT resource capacity is added in anticipation of demand. This strategy can take advantage of volume purchases but is more susceptible to over-provisioning and can result in unused resources that become obsolete before use.

Mainframe Early computers that were gigantic, not too reliable, and used large amounts of power. The first systems were dedicated to single users and were developed for specific business, government, or scientific tasks. IBM dominated this marketplace and gained fame as the computing platform for businesses.

Match Strategy Capacity planning approach that involves adding IT resources in small increments in an attempt to match demand. This strategy avoids over-provisioning but sometimes fails to take advantage of volume discounts.

Middleware The software layer between operating system and applications on each side of a distributed computing system. Common middleware applications include web servers, application servers, content management systems, databases, and tools that enable database integrations.

Operating System Virtualization Software permits hardware resources to run multiple operating system images simultaneously. These virtual operating systems appear real to the user even though they are abstracted away from the physical hardware.

Over-Provisioning Opposite of under-provisioning. Too much IT capacity exists, and the system is subject to waste and idle resources.

PaaS Approach Related to SaaS but is a broader and more complex form of cloud computing generally used by organizations involved in custom application development. Usually, PaaS solutions make it easier to develop, test, collaborate, maintain, track releases and changes, and perform other duties related to software creation. PaaS generally offers a configured sandbox for software testing, project management features, and a deployment environment that enables customers to roll out their cloud-based applications.

Platform as a Service (PaaS) Hardware and operating system are supplied and managed by an external party. Hardware functions are mostly transparent to the user.

Private Cloud These clouds are hosted on hardware systems located within an organization's data center. A private cloud is essentially an intranet where all the data and services are protected by a firewall. Also called an internal or enterprise cloud.

Public Cloud Turns day-to-day management of hosting to a third-party vendor. Generally, the organization using the public cloud is not responsible for managing the system, updating software, and maintaining the hardware. Instead, data and applications are stored in the host's data center where all IT operations are managed.

Recovery as a Service (RaaS) Includes cloud services that facilitate backup, archives, disaster recovery, and business continuity functions. RaaS ensures an organization has data backed up in multiple locations and can quickly resume operations should a natural disaster or other unexpected event occur.

SaaS Approach This approach to cloud computing centers on a software vendor hosting application software at their data center and, in most cases, a customer accesses it via a Web browser or a custom interface installed on the user's computer.

Scalability When associated with computing solutions, this means a system can grow without being hampered by existing structure or available resources. A scalable system can respond to higher demand with little or no changes required.

Service Oriented Architecture (SOA) A philosophy that suggests large development projects can be organized through a divide and conquer and communicate approach. As such, SOA describes both an architectural style and vision of how an organization should approach developing, building, and deploying systems. The main mindset of SOA is to develop reusable services integrated to create large scale systems.

Software as a Service (SaaS) Hardware, operating system, and applications are supplied and managed by an external party. Everything but the applications are transparent to the user.

Software Stack The set of software components required to complete an entire application. This means no further components are needed to build a complete software system. Sometimes called a Solution Stack.

Solution Stack The set of software components required to complete an entire application. Sometimes this term is used interchangeably Software Stack but can be different because it may include hardware components in the overall solution. In cloud-based environments, these have become virtualized and often integrated into PaaS cloud models.

Thick-Client Computing Applications run on desktop computers instead of from shared, centralized computing sources. Computing power resides on the client's computer.

Thin-Client Computing Uses applications running on remote processors. Computing power resides on shared resources rather than the client.

Under-Provisioning Under-provisioned systems underestimate resource demand and cannot meet the needs of the users.

Virtualization Originally used in mainframe environments, this is the process of commoditizing hardware, software, or other resources by creating representations that serve multiple users. The idea is to run an instance of the resource in a layer abstracted from the physical hardware.

Chapter 2 List of Terms

Accounting Software Newer technologies enable software vendors to provide cloud-based solutions for keeping organizational, financial records. Functions include accounts payable, accounts receivable, payroll, general ledger, financial reporting, maintaining controls, annual tax returns, and other processes.

Azure Microsoft's PaaS and IaaS cloud offering. It provides many cloud infrastructure and platform options.

Customer Relationship Management (CRM) SaaS applications enable organizations to interact with customers in ways that are tracked and maintained. The software helps build a relationship with a client and ensure the relationship is reinforced positively through each contact, even if different organizational members make the contact. CRMs keep customers connected and provide a consistent message to a client from all parts of the organization.

Data Analytics The art and science of transforming, collating, and analyzing raw *data* to draw conclusions about the resulting information. Many software packages exist to support these processes.

DevOps A combination of philosophies, practices, and cloud-based tools used to enhance an organization's capability to design, build, configure, and rapidly deliver applications and services. DevOps seeks to improve products at an accelerated pace. It is an evolutionary step past traditional systems development and IT infrastructure management.

Enterprise Resource Planning (ERP) A suite of business tools used to automate all basic business functions. Generally, ERP software integrates applications using a consolidated database and ties all business functions to an accounting system. Typical items integrated into an ERP system are human resources, CRM, invoicing, supply chain, inventory control, finance, project management, procurement, accounting, sales, and many other business functions.

Financial Accounting Accounting practices focused on collecting and reporting information for external users like investors, creditors, regulators, bankers, and auditors.

Help Desk and Service Software SaaS technologies enable moving customer service software features into the cloud. This software is useful for organizations or IT departments that support many users. Essentially it uses software to track, manage, and resolve customer issues.

Home Media Server Software to provide storage on existing computer hardware. Can be configured to enable access to music, video clips, movies, and other media from networked computers or hard drives. Clients include mobile devices and networked home computers.

Hosted Personal Cloud Storage Using Third Party Hardware An option for creating a personal cloud using third-party hosting services where cloud software can be installed and deployed.

Human Resources (HR) Software SaaS applications enable organizations to manage processes related to their human capital in the cloud. This includes many routine employee interactions such as storing and updating employee data, payroll, vacation tracking, sick or personal day tracking, training records, certification, recruitment, benefits administration, attendance records, evaluation, legal requirements, and performance management tracking.

Managerial Accounting Accounting practices focused on data collection and processing into information reported to internal managers and executives to support planning, controlling, and decision-making.

Microsoft Office 365 A hosted, cloud-based SaaS solution that frees users from installing MS-Office software on their own in-house servers. Office 365 is subscription-based and includes Office, Exchange Online, SharePoint Online, Lync Online, and Microsoft Office Web Apps. An organization using 365 can administer users from an online portal to set up new company user accounts, control who can access various applications and features, and see current operating statuses of online tools and services.

Microsoft Power BI This is Microsoft's primary data analytics software package featuring excellent graphics and visualizations.

Office Automation Software SaaS technologies enable automating routine business office tasks and creating items such as various documents, spreadsheets, calendars, and presentations. This software usually permits greater levels of sharing and collaboration since it is cloud-based.

Private Cloud Storage (PCS) A cloud storage model where users do the same things that public cloud subscription services do, including syncing devices, sharing files, and accessing content from smart phones or other devices. A big difference exists: the individual owns the hardware.

Public Cloud Storage Service Inexpensive (often free for smaller amounts of storage) storage generally found in SaaS cloud models. Example services include Box.net, Dropbox, Google Drive, Carbonite, and Apple iCloud. In most instances, the user creates an account, logs in, and uploads files where they remain until the user decides to download them for later use.

Purchasing and Procurement SaaS technologies enable moving purchasing, procurement, and vendor management processes into the cloud. Software supporting this area helps organizations find, buy, and manage supplies and services required to conduct business. A wide range of software exists in this area.

Redundant Array of Independent Disks (RAID) A technology that redundantly stores data in different places on multiple hard disks to protect data in case of a drive failure. Essentially it is a safety measure internal to the storage device in case part of it fails.

Small and Medium Enterprises (SME) Small and medium businesses are described as SMEs.

Social Media Interactive, cloud-based software applications used to enable user connections through various communication channels. Typically facilitates creation or sharing of information, ideas, commentary, and expressions via virtual networks and online communities.

Social Media Strategies A planned, long-term approach to using social media to enhance business operations. Engaging the target audience is key to understanding what platforms and content work best to connect and engage key individuals.

Tableau Online A data analytics software package hosted in the cloud which allows users to upload data, publish dashboards, and share reports via a cloud interface. Specializes in data visualization and the creation of graphic output to represent data.

Tableau Public A free, cloud-based tool with most Tableau Desktop features, except it does not permit users to acquire data from as many sources. The results are shared publicly on the cloud which means any mission critical or sensitive data would not be a good fit for this public solution.

Ticket Management Tickets are customer requests for help used in help desk software. These requests are filtered and categorized for assignment to team members, prioritized for urgency, and maintained with customer history. Tickets can be viewed by all relevant people and ensure everyone is on the same page regarding resolution and customer communication.

Chapter 3 List of Terms

Bare-Metal Virtualization Installation of a hypervisor as an operating system. In this instance, the hypervisor or VMM is installed directly on computing hardware.

Container Provides virtual space for applications which do not experience any difference from running solo on a physical machine. Applications have access to all permissible resources such as disk drives, files, network connections, and hardware items. Containers enable applications to run in efficient isolation on a shared operating system kernel without needing the deployment of VMs. Containers share the host's OS.

Container-Based Virtualization An approach to virtualization where the abstraction layer is an application on the operating system. Isolated guest virtual machines, called containers, are installed in the application.

Containerization Similar to operating system virtualization but uses a host operating system feature where the kernel permits creation of isolated, user-space instances called containers. Also called Operating System Level Virtualization.

Containers as a Service (CaaS) Cloud-based service model where containers and related applications are uploaded, organized, and managed to facilitate use of this technology.

Daemon Docker's persistent process to manage user-defined containers and container objects. The process is called dockerd and listens for requests sent to the Docker API.

Direct Attached Storage (DAS) Computer storage devices directly attached to one computer. DAS is not meant to be networked and therefore does not make good cloud storage. Instead, it is intended for individual computer users and takes the form of a hard drive or a solid-state drive (SSD). From an enterprise computing standpoint, DAS can include a server's internal disk drives and external drives attached with SCSI or other interfaces. In this case, the drives may be accessible via a network but only through the server to which they are attached.

Docker A platform for building, distributing, deploying, and managing containers.

Dockerd Docker's daemon used to listen for requests sent to the Docker API.

Docker Containers Standardized, encapsulated environments used to run applications. A container instance is managed through the Docker API.

Docker Cloud A native solution used to deploy, monitor, and manage Dockerized applications.

Docker Hub Docker's default public registry.

Docker Images Read-only templates used to build the contents within a container. Images deploy applications after testing and debugging, usually in multiple locations.

Docker Registries Docker's repository for retaining images. Clients can connect to registries to allow images to be downloaded (or uploaded). Docker offers both public and private registries.

Docker Services Permit scaling containers across multiple Docker daemons. A swarm of cooperating daemons are created, and these daemons communicate through Docker's API.

Docker Swarm An orchestration tool used in conjunction with Docker containers.

Fibre Channel (FC) High speed device communication protocol that overcomes the speed constraints of iSCSI and enables high speed data transfer. Like iSCSI, fibre channel is used to connect storage devices using many different approaches.

Full Virtualization Emulates the full set of hardware needed to run a VM in a hosted environment. A fully virtualized instance runs on top of a guest OS. The guest OS runs on the hypervisor which runs on the host's operating system and hardware platform. This form of virtualization is meant to be indistinguishable (from an application standpoint) from an actual hardware platform.

Graphics Virtualization Technology A popular hardware virtualization related to graphics technology. It revolutionizes video game speeds and makes it possible for video-related processing to be moved off the main CPU.

Hardware Assisted Virtualization Uses the host computer's existing hardware to build a VM.

Hardware Virtualization Using virtualization techniques and software to create multiple virtual machines on an existing, physical computing platform.

Host Virtualization An alternative to bare-metal virtualization where the OS is installed first then standalone hypervisor software is installed. VMs are created on this platform. This approach has fallen into disuse due to integration of many OSs with hypervisor functionality. However, in recent years, container-based virtualization, using the same ideas, has become popular.

Hyper-V and Windows Server Containers Microsoft's container software. Hyper-V containers use a small footprint VM to provide application isolation.

Hypercalls Calls between guest OSs and hypervisor.

Hypervisor Creates, runs, and monitors virtual machines. A hypervisor most commonly is deployed as software but can be firmware or hardware. The hypervisor runs on a host machine housing the virtual machines, sometimes called guest machines. The hypervisor's primary functions are to ensure guest machines have a virtual hardware platform and then manage requests from guest machines' operating systems in interactions with the virtual hardware platform. Hypervisors permit multiple instances of operating systems to operate on isolated VMs simultaneously. Also called a Virtual Machine Manager (VMM).

Image Comprises software needed to run a configuration on a computer system. Often includes an OS, executable files, and data files needed to run a system. In a sense, a filesystem having both directory structure and necessary files. An image allows a computer to be restored to the same state in multiple instances.

Internet Small Computer System Interface (iSCSI) Device communication protocol that moves SCSI onto the Internet using an IP-based standard for connecting servers to storage devices.

Kernel Centerpiece of a computer's operating system. This program controls all other programs. The kernel provides services that enable other programs to request various hardware items central to the computer. It also sends requests to device drivers that control attached hardware. The kernel regulates CPU use and enables multiple programs to use it simultaneously (e.g. multitasking).

Kubernetes An open-source, former Google product that provides a portable, extensible, open-source platform for organizing containers into pods located on nodes. Kubernetes is used for many container functions and helps automate, deploy, scale, maintain, and operate application containers. Kubernetes is considered an orchestration tool in many of its uses.

Microservice Architecture Software development approach where an application is broken into smaller, specialized parts that communicate across REST-based interfaces or APIs. In general, microservices manage their own data, authenticate, and maintain logs. Microservices are ideal for container environments.

NAS Head A protocol handler gateway specifically engineered to interface with a SAN. A NAS head does not have any client storage availability, instead it provides a NAS-style interface that, from the end-user's perspective, makes storage appear to occur at the file-level. The NAS head offers the load-balancing and protocol handling needed to transparently change file-level access to more efficient, block-level format access.

NAS Virtualization In a sense, this is the creation of a private cloud for storage. Software products exist to enable an existing NAS device to abstract stored files from their physical location to an online virtual location.

Network Attached Storage (NAS) A form of storage configuration. NAS uses a dedicated piece of hardware, called an appliance, to add commonly accessible storage to a network. In larger, complex systems, a NAS appliance can help manage file requests, provide additional security, and offer methods for configuring and overseeing a shared storage system. Another unique feature of NAS is its use of file level transfers as opposed to the block level approach more commonly used in DAS and SAN approaches. Uses an existing network for the appliance and the storage devices.

Operating System Level Virtualization Like operating system virtualization but uses a host operating system feature where the kernel permits creation of isolated, user-space instances called containers. Also called Containerization.

Operating System Virtualization A technique where VM software is installed on the host computer's operating system. This contrasts with bare-metal virtualization.

Paravirtualization This virtualization approach requires guest operating systems to be modified to work on the virtual machines running on the host. Hardware is not exactly emulated. To counter this, the VMM uses an application program interface (API) approach. It modifies the guest machines' operating systems and replaces the existing code with instructions to enable calls to the VMM's APIs.

Processor Command Set Extensions Hardware features added to a CPU's processing capabilities. Developed by chip manufacturers to enhance speed performance in microchips likely to be used in servers hosting VMs.

Serially Attached SCSI (SAS) Like SCSI, this device communication protocol uses the same basic command set to share data with attached devices. SAS uses a serial connection to increase throughput and transfer data at higher speeds. SAS is a full duplex technology which means it can send and receive data simultaneously.

Small Computer System Interface (SCSI) A standard protocol to ensure devices can speak the "same language." Many hard drives, CDs, DVDs, and other storage devices are attached using SCSI. SCSI's parallel interfaces require use of a microprocessor-controlled, smart bus, that permits peripheral hardware devices to be daisy-chained to a computer. In addition to storage devices, other hardware such as scanners, printers, and others use SCSI.

Storage Area Networks (SAN) A form of storage configuration that comprises switches, disk controllers, host bus adapters (HBAs), and fibre or iSCSI connections to enable users to share storage arrays in ways that are flexible, scalable, and ensure data redundancy. In general, SANs provide the ability to share storage arrays among application servers.

Virtual Machines (VM) Emulate a hardware system meaning multiple OSs can run in independent environments on the same physical machine. Hypervisor software emulates the hardware required from pooled CPUs, memory, storage, and network resources. The hypervisor ensures resources can be shared by multiple VMs. Therefore, VMs take more memory since each one requires a guest OS.

Virtual Machine Manager (VMM) Software installed on a physical machine used to create, run, and monitor virtual machines. Utilization of the physical hardware (e.g. processor, memory, and other hardware resources) is tracked. Also called a hypervisor.

Virtualization A logical (e.g. virtual), abstracted version of a resource, such as a storage device, server, computing desktop, operating system, or network resources that can be accessed by users transparently and without added effort.

Chapter 4 List of Terms

Ant Colony Load Balancing Algorithm An imaginative load balancing method based on the natural world. It views the server pool from the perspective of worker ants. Ants are essentially intelligent agents that move forward and backward to track overloaded and underloaded nodes. Ants update "pheromones" to track node resource information. Two update types are: (i) foraging pheromones which help find overloaded nodes; and, (ii) trailing pheromones to help find underloaded nodes.

Autoscaling A cloud computing best practice for managing elasticity. Existing servers are watched or monitored and when an important usage threshold is reached, the system expands and adds a server. Autoscaling can be based on business rules. On-demand cloud computing's autoscaling feature sometimes is called elasticity.

Back-End Services Software that remains transparent to the user and interfaces with system resources.

Bee's life Load Balancing Algorithm This algorithm is inspired by the natural world and a honeybee's process of searching for food. The bee manages loads across VMs in ways that reduce task queue waiting times.

Cattle Servers Uniform servers characterized with standardized features. That makes these servers easy to duplicate and replace without extra setup time.

Central Load Balancing Algorithm This approach, as the name implies, uses a centralized and dynamic approach. When the central manager receives a request, it is stored in a FIFO (first-in-first-out) queue. The requests remain in the queue until a processing node sends notification that it needs more work. The oldest request is removed from the queue and work is sent to the node. If no work is in the queue, the request for more work is stored until a request comes.

Central Manager Load Balancing Algorithm This balancing algorithm is used in the cloud as well as in traditional client-server networked systems. A central server selects a node for load transfer. Generally, the server having the least current load would be chosen to receive the newest request.

Client-side Load Balancing in the Cloud In this load balancing approach, middleware acts as a client for back-end services and uses an elastic cloud storage service to choose back-end database or web servers based on pre-determined criteria like current load, queue length, average processing time or other factors. In this case, the client finds the best service instead of a central manager or agent doing it.

DNS Balancing A network optimization technique used to route a web domain's incoming traffic flows to an appropriate web server. DNS load balancing is concerned with faster access to resources meant to be served back to a client connected to a website. DNS load balancing distributes load requests for a domain.

Dynamic Balancing Load-balancing approach that uses current load information for request distribution while the system is running. Dynamic algorithms are adaptive and change as the system state changes. Unpredictable, varying loads work well with a decentralized, dynamic algorithm.

Elasticity Methods used to adjust cloud resources on-the-fly to match workload changes. This can be up or down in size.

Front-End Services Software that interfaces with the user.

Genetic Algorithm Load Balancing Algorithm This natural world motivated algorithm uses a genetic algorithm approach to optimize node selection process. This algorithm employs a "survival-of-the-fittest" paradigm to ensure algorithms with higher efficiency produce "offspring" that may inherit their best features over a series of generations.

Hybrid Load Balancing Algorithm Many hybrid load balancing algorithms exist. In general, these algorithms use multiple stage approaches to determine the best node to receive a processing task.

Infrastructure as Code (IaC) Self-documenting process that ensures anyone with programming or scripting knowledge can understand system setup and infrastructure. IaC uses scripting languages such as XML, JSON, or Python to document system configuration.

IP Hash Load Balancing Algorithm This algorithm is very similar to the random method. However, instead of selecting the next server randomly, the IP address of the client is used to select the host to service the request. This distributes load somewhat randomly.

Latency Response time regarding computing requests.

Load Balancing Process of efficiently distributing incoming network traffic among servers or nodes.

Least Connection Load Balancing Algorithm This algorithm looks at load distribution based on the number of active connections between a server and its clients. Distribution of new loads (or re-distribution of tasks) is influenced by rules for keeping the number of connections approximately even.

Middleware Software that acts as a bridge between backend services and users' applications.

NGINX Plus This load-balancing software, used by many leading cloud-based organizations, helps with processing requests. High-volume sites like Netflix and

Dropbox have used this cloud load balancing solution to ensure content delivery in secure, reliable fashion.

Noisy Neighbors Clients that monopolize bandwidth, storage access, CPU, and other resources that impact other clients in multi-tenant cloud architecture.

Pay-as-You-Go The practice of paying only for the computing, storage, and networking resources that an organization uses.

Pet Servers Highly customized servers with hard-coded processes and specialized features. Pet servers are almost impossible to replace quickly if something goes wrong.

Random Load Balancing Algorithm Randomly selecting the next server to receive a request is one static method used to distribute workload. This approach does not require any information about the last server selected nor does it attempt to determine the current loads on each server. Random selection means that over time every server will receive approximately the same number of requests.

Round Robin Load Balancing Algorithm This static approach to node selection is best suited when servers are homogeneous, and requests are approximately the same magnitude. In this scenario, server assignments are put into an ordered sequence.

Scalability A system's capability to accommodate larger loads easily by adding resources.

Scaling Out Duplication and expansion by adding more resources. A second server may be added to increase capabilities.

Scaling Up Increasing individual capabilities. A server may be scaled up by adding more memory or more storage.

Static Load Balancing Load-balancing approach that depends on a set of rules when a processing node is selected. In general, systems expecting constant, uniform requests work well with centralized, static load balancers.

Threshold Algorithm Load Balancing Algorithm This algorithm creates new nodes as demand increases. Loads are assigned immediately to these new nodes.

Transfer Load Balancing Algorithm This algorithm looks at ongoing server loads and determines if one node is overworked. If so, the central manager can decide to move a task from one server to another for processing. Basically, this algorithm was created to manage inter-process task migration to help keep all nodes operating at relatively similar levels.

Virtual Local Area Networks (VLANs) Portions of a network partitioned into components which match the business needs of those using computing resources. For instance, the VLAN might be partitioned by functional business department, or by required security levels. VLANs can divide a system into logical groups and establish related rules about use. VLANs can be simplistic and mimic a traditional office environment with printers, shared drives, and so forth. Or, they can be highly complicated to meet security clearance and legal requirements of a business or industry. A VLAN can also mimic geographic configurations if that has an advantage.

Chapter 5 List of Terms

3-2-1 Rule Best practice for data safety that suggests that an organization should always maintain (3) copies of its data, back up its data using (2) different forms of storage, and always keep at least (1) copy of its data offsite.

Advanced Encryption Standard (AES) Technology developed for securing secret government documents. Often, storage vendors use this technology so neither they nor any intruders can access client data.

Application Programming Interface (API) An interface that allows one piece of software to talk to another. Essentially, it is a translator that takes the conventions of one system and transforms it to the conventions of another.

Automation Replacing manual tasks with automated solutions.

Automation Task A specific element of automation.

AWS CloudFormation A configuration and orchestration tool developed and released by Amazon that permits IT specialists to codify infrastructure and automate cloud resource deployments. CloudFormation relies on templates which use YAML, JSON, or are visually developed.

AWS CloudFormation Designer A drag and drop environment for visually developing AWS Cloud formation templates.

Azure Resource Manager Microsoft's native infrastructure management tool for Azure cloud. The tool considers the complex environment that Azure may be asked to deliver replete with VMs, networks, storage, databases, webservers, microservices, third-party software installations, and so forth.

Business Continuity (BC) Processes and activities that ensure business operations remain as normal as possible under all conditions, particularly following catastrophic events.

Chef This is an automation framework widely used to setup, configure, deploy, update, manage, and remove servers and applications in a range of environments including cloud-based ones.

Chef-Client An agent running on every node to assist with Chef-server interaction.

Chef Cookbook Contains many recipes and defines scenarios to automate installation and deployment of resources. Think of cookbooks as an orchestration of recipes.

Chef Infra An automation platform used to transform infrastructure into code.

Chef Node Any physical, virtual, or cloud machine served by the Chef system.

Chef Recipes Define resource configurations and are used to install, update, and deploy IT infrastructure elements.

Chef-Server Central controller for the Chef system that serves out recipes to nodes.

Chef Workstation Station where Chef recipes are developed and deployed to the Chef server for use by the nodes.

Cloud Automation Catch-all phrase used to describe the processes and tools that replace manual tasks related to managing organizational computing infrastructure.

Cloud Backup Processes that store resources at points in time so systems can be returned to prior states should the need arise.

Cloud Replication A scheme that protects IT infrastructure and data against catastrophic events or disasters by continually replicating (this means creating an exact copy) data and/or VMs, servers, and other virtual resources in diverse locations. If a failure occurs, a working version of the system can be started in a new location quickly.

Data Replication Process of creating exact copies of data to enable more uses of existing data without interfering with critical operations. It helps ensure data security.

Disaster Recovery as a Service (DRaaS) A cloud-based, elastic service used to quickly restart an organization's infrastructure should a catastrophe occur. Virtual machines, data, containers, and so forth all have a virtual footprint that has been duplicated and are put back into motion very quickly.

Environment Drift The result of sprinkling custom features into an IT environment deployment. IaC prevents this situation from emerging.

Facter Puppet's utility program that discovers existing system information and creates a catalog with all resources and dependencies.

Google Cloud Deployment Manager Used to facilitate IT shop needs for declaring, configuring, deploying, and managing cloud resources. It uses JSON or YAML to describe deployment rules.

Idempotent Describes algebraic elements that do not change when raised to a positive integer power. In cloud computing, this term describes the goal of IaC: ensuring that code produces the exact outcome each time it is used.

Infrastructure as Code (IaC) Using program scripts to automate mundane system configuration and other IT-related tasks. This helps organize best practices and prevent the operations team from reinventing the way to bring up servers, perform backups, configure desktops, and many other tasks.

Locking Database management features that prevent accidental removal of key resources without taking extraordinary measures.

Orchestration A type of automation that coordinates and organizes multiple automation tasks.

Permission Set A mechanism used in role-based access control. A group of resource and/or action permissions is given a name and is used in multiple assignments.

Pull Approach Client resources contact a central server to request infrastructure updates. The central administrator has less control over configuration of resources. This approach often is used in places where client users have the expertise to manage their own systems and need to manage their current configuration.

Puppet An open-source configuration management tool that uses declarative language to describe desired system configuration. It works in virtual and cloud-based systems.

Puppet Enterprise An enterprise edition of Puppet that provides tools to manage configurations and support migration from physical data centers to cloud-based solutions.

Puppet Manifests Templates used in Puppet configurations. The manifests are created with Puppet's declarative language or Ruby commands.

Push Approach Uses a centralized server to push out infrastructure changes. It allows administrators to carefully control production environment changes and deployments.

Replication Creates an exact copy of systems, data, configurations, and other IT resources to ensure data protection, integrity, and transaction speeds. Replication may occur because of application migration processes, data warehousing, data mining, and analytics operations.

Role Assignments A role-based access mechanism used to control who has permission to conduct specific actions regarding system resources. This approach attaches a definition to a user or user group. Users generally are assigned to predefined roles. The roles may offer different users or individuals various levels of access and control over resources.

Role Definitions A role-based access mechanism used to control who has permission to conduct specific actions regarding system resources. This approach is based on people's roles in an organization. Role definitions describe a permission set. This set is given a name and then can be used in multiple assignments. The definitions can be hierarchical and inherit characteristics from other definitions.

Salt Another name for SaltStack.

SaltStack A Python-based open-source software system that supports the IaC approach to enterprise IT resource management. SaltStack is considered by many experts as the most scalable, best system for very large networks of servers. Also called Salt.

Secure Socket Layer (SSL) A form of data security that focuses on encryption during transmission. Essentially it ensures data cannot be deciphered if a wire is tapped or if data is intercepted wirelessly.

Snowflake A unique IT environment configuration caused by environment drift. This configuration cannot be reproduced automatically and results in inconsistencies among deployments. Administration and maintenance of snowflake infrastructure involves manual processes which are hard to track and ultimately contribute to errors.

Software Development Kits (SDKs) Toolkits that permit IT specialists to use a software package or resources on their platform. SDKs generally include tools, samples, pre-written code, processes, documentation, guides, and other resources to make implementation easy.

System Latency Time required to restart in the event of a catastrophic IT system failure.

Terraform An IaC tool used to create, change, and update IT resources in cloud environments. It uses a series of code-based configuration files to drive infrastructure deployment. It uses a datacenter paradigm and declarative approach to help structure cloud-based environments.

YAML A markup language with data-oriented use and intention. The acronym YAML originally meant, "Yet Another Markup Language." Later, the developers suggested its name was recursive and really stood for: YAML Ain't Markup Language.

Chapter 6 List of Terms

Amazon CloudWatch A native suite of tools provided by Amazon used to oversee, monitor, and diagnose issues on AWS Cloud. This system is included and available to all AWS Cloud users.

Cloudability Third party billing tools that focus on cloud cost and usage data for organizational decision-makers. Provides specialized reports focused on operations, accounting needs, and management team.

Data Breaches Situations where hackers illegally obtain data from a cloud provider or other source.

Data Disposition Clauses Part of an SLA that deals with data ownership after a contract is terminated. Generally, it covers moving data to a new provider and/or removing it from the prior cloud vendor's custody.

Data Location Physical location where data reside in a cloud environment. This is important since the laws of countries and jurisdictions vary, and legal problems concerning data migration, import, export, and storage can result.

Disclosure Clauses Part of an SLA that specifies how and when a vendor is required to disclose details when a data breach occurs.

Earn-backs A cloud vendor's right to reduce service credits or penalties due to cloud environment performance issues.

Google Cloud Operations In 2014, Google acquired Stackdriver, a provider of cloud management software which generated performance and diagnostics data derived from operations in Google Cloud. The software was renamed Google Cloud Operations. It also operates in AWS, making it a hybrid solution. In general, Cloud Operations offers similar features to Azure's and Amazon's native monitoring tools with monitoring, event logs, activity traces, error reporting, and alerts.

Indemnity Clause Important aspect of the SLA for the cloud service client. Protects against lawsuits caused by cloud vendor actions.

Key Business Performance Metrics These cloud performance measures consider key business performance indexes provided by a cloud vendor.

Microsoft Azure Monitor Microsoft offers a suite of monitoring tools to help IT administrators gage the performance of their cloud-based resources. The tool set is built into their Azure cloud infrastructure and using it just requires activating the service.

Security Metrics These metrics assess cloud performance for organizations concerned with privacy and security-related issues.

Service Availability Metrics Common metrics used to assess vendor obligations in cloud environments. Vendor guarantees uptime of system to a contractual level.

Service Performance Metrics Cloud performance metrics that look at items like how long database reads or writes take; time required to provision a new VM; how many retries a database attempt gets; and, so forth. Often, these are custom metrics closely aligned with business goals of the cloud customer.

Service Quality Metrics Cloud performance metrics that assess the accuracy of storage, backups, web services, and other items. Defects, failures, or other measures of quality often are considered.

Single Pane of Glass Monitoring Solution The idea is that an enterprise should have one monitor that comprehensively groups all metrics into a single interface and provides drill-down capabilities for that tool.

Service-Level Agreement (SLA) A binding contact between a cloud vendor and their client. Generally, the cloud vendor is an organization like Microsoft or Amazon, and the client is a business organization moving their IT infrastructure into the cloud. The SLA describes conditions for use and minimum level of service and performance the cloud vendor will provide. Among common service and performance levels are uptime, availability, reliability, and responsiveness.

SLA Data Ownership Clause This is a primary SLA clause that specifies cloud data ownership and management under a variety of circumstances.

SLA Metrics Specific, measurable benchmarks to define what constitutes high quality service outcomes in cloud environments. The terms are defined prior to signing an SLA to remove any doubt regarding expectations of both parties.

Zombies The living dead just like on the movies.

Zombie Instance These are resources that may have once been useful, functioning members of the IT resource pool but are now hanging around, consuming costs without any contribution.

Chapter 7 List of Terms

Amazon Cognito This is Amazon's federated identity pool that enables users with identities from various providers to access their resources. Cognito is intended to help users working in one of Amazon's clouds, often from a mobile platform. It allows an IT administrator to create unique identities with customized permissions or to permit users authenticated elsewhere to enter Amazon resources.

Business Administrator Generally an IT staff member who oversees the process of developing a list of resources and template for matching individuals to those resources in cloud environments.

Cloud Access Security Broker (CASB) Cloud security features offered by PaaS vendors that ensure visibility, compliance, data security, and threat protection.

Delegated Access Clients (which are applications) access resources and take actions for the user, without the user ever sharing their personal information including ID and password with the application.

Dormant Resources Resources taken offline temporarily. If a security update occurs during that time, it is possible the item is out of date and vulnerable to attack. Also called temporary resources.

Federated ID Management (FIM) Provides users with one set of credentials to access all organizational computing resources in a secure and trustworthy manner and provides users with access across different organizations or cloud systems. Also known as ID as a Service (IDaaS).

Flow The process of using delegated access to seamlessly move from one cloud resource to another.

Google Cloud Identity and Access Management (IAM) Google offers this service to give administrators the ability to determine which users can view, change, or access various cloud-based resources. It enables security policy development and enforcement. It offers auditing and reporting service capabilities as well. Cloud IAM allows users to access Google-provided services and tools (e.g. Docs, Gmail, Sheets, and so forth) but does not perform user authentication for services outside of Google.

ID as a Service (IDaaS) Cloud service from an ID provider such as Google, Microsoft, or Okta. Used to provide user ID and password functionally for cloud applications. Also known as Federated ID Management (FIM) systems.

Identity and Access Management (IAM) Security practices that manage policies to determine who can access various cloud resources.

Identity Provisioning Securely adding and removing cloud users in timely and effective ways.

Leakage When information that is hardcoded into computer code, including sensitive items such as paths, userIDs, passwords, and other secure artifacts in human readable form are unintentionally dispersed.

Microsoft Active Directory (AD) Windows OS service used to connect users with network resources. It offers an interface to give IT Administrators access control to network directories, folders, and resources. It controls user access based on security policies set up and organized by an IT administrator.

Microsoft Azure Active Directory (Azure AD) Builds on Microsoft's Active Directory (AD) services to provide multi-tenant, cloud-based identity management services. This FIM focuses on access management, services provided by Microsoft AD, and identity protection across multiple platforms.

Network Packet Brokers (NPB) Devices that offer monitoring capabilities and tools to oversee network traffic at cloud endpoints.

Okta This is an identity management solution that works across many emerging major platforms in cloud computing. It uses an organization's on-premise or cloud identity management system and integrates that with its cloud-based tool.

Open Authorization (OAuth2) An open standard used for authentication and authorization with Internet-based systems (although it is not limited to this). OAuth2 is token-based which means after a user enters a correct user ID and password, they are given a token which enables them to access a resource. The token remains valid for a pre-determined time and may also be accepted by other resources.

Orphan Storage Like zombie servers, orphan storage can be remnants from testing or development that have not been deprovisioned. Storage issues also relate to disks not attached to the system in a formal way. If critical data were placed onto an unmonitored location, leakage or loss can occur. Also called Zombie storage.

Per Use Cloud software licensing model that monitors either the numbers of accesses, the amount of time spent in application, or other usage metric to determine price.

Per User Cloud software licensing model assesses total user numbers to provide a price. Per user approaches may be based on seats, meaning a hard limit on the number of simultaneous users at any given time.

Resource Misconfiguration Common mistakes made related to how and where various cloud systems are deployed. This can happen when scripts are created for automation or orchestration. Certain actions make a cloud resource more vulnerable to error, attacks, or other problems.

Security Assertion Markup Language (SAML) An open source standard for exchanging authentication and authorization information between identity providers and service providers.

Shibboleth A widely used, open-source FIM often deployed in higher education environments. It allows SSO within organizations and permits federated logins across enterprises and domains. It also permits sites to protect online resources and ensure privacy of critical data.

Single Sign On (SSO) Provides users with one set of credentials to access all organizational computing resources in a secure and trustworthy manner.

Temporary Resources Resources taken offline temporarily. If a security update occurs during that time, it is possible the item is out of date and vulnerable to attack. Also called Dormant Resources.

Universal Directory (UD) This tool from Okta stores relevant information about users and their cloud resource permissions.

Visibility Ability to see resources in cloud computing. An organization loses system visibility when it uses PaaS or other cloud infrastructure because the cloud provider manages most aspects of resource usage and user access.

Workload Problems Hackers can use cloud resources to set up mail servers, or programs that attempt to launch attacks or determine passwords. These are essentially rogue processes that should not be running.

Zombie Servers Servers or related resources started as tests or for other reasons that are not deprovisioned afterwards. These instances can be security risks because they provide entry points into resources that may not be watched carefully.

Zombie Storage Like zombie servers, zombie storage can be remnants from testing or development that have not been deprovisioned. Storage issues also relate to disks not attached to the system in a formal way. If critical data were placed onto an unmonitored location, leakage or loss can occur. Also called orphan storage.

Zombie Workloads Resources started as tests or for other reasons and are not deprovisioned afterwards. These instances can be security risks because they provide entry points into resources that may not be watched carefully.

Chapter 8 List of Terms

256 Bit Encryption Strong form of encryption commonly used for data being transmitted from one server to another. The 256-bit part refers to the encryption key's length. To break this encryption, a hacker would need to try up to 2^{256} combinations before absolutely finding the correct outcome.

Acceptable Risk IT auditors use this term to consider cost and probability of occurrence, then they develop a smart solution with reasonable protection against critical problems rather than eliminate all risk.

Application Controls IT controls that seek to build robustness into cloud systems. They include simple examples such as login and password policies, and account lockouts after three failed attempts.

AS 8015-2015 This IT governance standard came about in response to the Australian Government's finding that poor corporate and information governance caused many business failures during the dot com crash. It provides principles, common language, and best practice models for corporate information and communications technology governance.

Capability Maturity Model Integration (CMMI) This IT maturity model was developed by the Software Engineering Institute to classify where on a scale from 1 to 5 an organization currently operates in terms of IT performance level, quality, and profitability. The current state of organizational IT drives process improvements required to get to the next level. Several best practices and measurements are recommended. The CMMI has limited applicability in cloud environments because vendors often supply mature solutions.

Certificate Authority (CA) An independent organization that issues digital certificates to ensure a web site belongs to the company that claims to represent it. The certificate authority seeks to stop middleman attacks or web site frauds where a site might pose as a legitimate company to obtain private information.

Chief Information Officer (CIO) Most senior technology position in an organization. The CIO is a strategic thinker charged with IT transparency, organizational and outside collaboration, and enhancing business value. Their role includes both overseeing operations and planning how IT can help enhance business opportunities leading to growth.

Cloud Encryption An effective way of securing data by scrambling files, data, or other stored items according to a complex pattern based on a key.

Cloud Governance Natural extension of the larger area of IT governance. In most organizations, no distinction is made between the two.

COBIT 2019 The gold standard IT governance framework. It offers a process for implementing IT governance within an organization from a business objective perspective.

Content-Limiting Filters A filter usually implemented to open a subset of web sites accessible by employees in their work environments. Filters can consider regulations, laws, and corporate concerns and limit social media sites or other places where data leakage might occur.

Corrective IT Controls IT controls that help "fix" problems.

Data Masking Form of protection which replaces sensitive data with fictional data that can be translated back at key times using algorithms built into software applications. This enables the data to be used for testing, data analytics, and other activities without exposing the contents unnecessarily.

Detective IT Controls IT controls that provide notification that a problem has occurred.

End-to-End Encryption Practice of encrypting data prior to transmission to the cloud where it remains encrypted until the client retrieves it.

End-User Controls IT controls that attempt to secure end-user computing resources such as mobile devices, social media, and cloud-based software subscriptions and prevent leakage of sensitive organizational data.

Factor Analysis of Information Risk (FAIR) This is an IT governance model that focuses on organizational risk. It was developed in the era of cybersecurity and cloud computing, so these aspects of risk informed the approach's development. FAIR helps organizations understand potential for loss and offers factors to preemptively stop problems.

General Controls IT controls to ensure correct access to resources, govern change management, and make sure physical security has been implemented. This means the right people have access to the right resources.

Hardware Security Module (HSM) A physical computing device implemented for on-premise encryption key management practices.

HSM as a Service Combines HSM capabilities with cloud service models to offer a high-level of security in multi-cloud environments for encryption key management. These systems are cloud-neutral and globally available with low latency.

Information Systems Audit and Control Association (ISACA) The organization that created and maintains COBIT standards. They provide a range of training programs, consulting references, and other services for IT oversight and auditing.

Information Technology Governance The portion of overall corporate governance dedicated to oversight of a firm's information technology activities, systems, assets, policies, and uses. It ensures that IT activities align with organizational strategies. Also called IT Governance.

Information Technology Infrastructure Library (ITIL) IT Governance framework originally developed within the British government in the 1980's to provide best practices information for its IT functions. The intention for ITIL was to help government operations standardize technology use and ensure IT supported core functions and needs. The framework uses a service-oriented perspective and has evolved into a general framework meant to support business operations.

ISO/IEC 38500:2015 IT governance standards that provide guidance to governance board members and those responsible for daily business operations. This standard was influenced heavily by AS 8105 and recommends clear delineation between management and governance to ensure oversight of sensitive issues. This standard also recognizes that ICT processes may be controlled either on-premise or by service providers (e.g. cloud vendors).

IT Auditors IT experts that examine whether an organization's relevant systems or business processes for achieving and monitoring compliance are effective. IT auditors assess the design effectiveness of rules and controls – whether they are suitably designed or sufficient in scope to properly mitigate the target risk or meet the intended objective.

IT Controls Policies, processes, and procedures used to mitigate known risks. An organization puts these into place to reduce risks that naturally occur in a process.

IT Dependent Controls These automate manual controls and provide system integration. For example, an IT dependent control may provide a list of all people that have logged into a system in the past month and send it to a supervisor for review. In general, IT dependent controls rely on some automation and some manual intervention.

IT Governance The portion of overall corporate governance dedicated to oversight of a firm's information technology activities, systems, assets, policies, and uses. It ensures that IT activities align with organizational strategies. Also called Information Technology Governance.

Key Management System (KMS) A software solution for securing and managing an organization's encryption keys. In this approach, keys are maintained in the cloud environment and secured using algorithms and best practice procedures.

Limited Encryption Practice of encrypting only the most sensitive data like passwords or customer credit cards numbers.

Manual Controls Controls outside the IT system that may use technology in their implementation. For example, things such as having a permission form signed by a supervisor, signing an Internet use policy after receiving training, or signing a travel reimbursement form indicting you agree with its accuracy.

Network Filtering Filtering implemented at the network level within an on-premise network or over virtual networks. This filtering can block data loss and ensure files do not leave the organization's secure areas.

Preventative IT Controls Most IT operations seek preventive controls. This means problems are stopped before they occur.

Registration Authority (RA) An independent organization in a network used to verify user requests for a digital certificate after which it informs the certificate authority (CA) to issue it.

Sarbanes-Oxley (SOX Section 404) A U.S. federal law that regulates financial aspects of a public firm with parallels in many countries. Its requirements impact IT audits since IT systems and cloud computing contribute to financial risk.

Secure Sockets Layer (SSL) A commonly used protocol for encryption in cloud computing that secures communications between a browser and web server. It is intended to reduce the chances that someone will intercept communications. SSL is transparent to users and this helps ensure its wide use.

Service Oriented Architecture (SOA) Governance Governance is put into place to ensure service quality, predictability, visibility, and cost-effective performance within an organization. It also ensures policies, laws, and regulations are followed by an organization. Its primary components are SOA Registry, SOA Policy, and SOA Testing.

Shadow IT Individual, departmental, or project team use of cloud-based applications from outside the formal IT infrastructure. In other words, all information technology applications managed by an organizational unit without formal knowledge of the IT department.

Sharding IT practice which breaks files into small chunks and then encrypts each separately. Each chunk is stored in a different location on the cloud provider's servers. If an intruder breaks through normal defenses and deciphers the encryption scheme, random chunks of meaningless data would be discovered.

SOA Policy Principles used to ensure services do not conflict and to ensure implementation follows good design, custom relationships, and compliant practices.

SOA Registry A catalog listing SOA services available. It can be used internally or as a tool to enable development of business partnerships.

SOA Testing A regular schedule of audits, tests, and performance metrics used to ensure operation of SOA. It intends to ensure SOA solutions are working correctly in a secure, cost-effective way. It also ensures regular system updates.

The FAIR Institute This is non-profit organization of risk specialists, IT cybersecurity leaders, and executives that maintain the FAIR governance model standards.

Threat Agents Practices that endanger IT-related assets in organizations. Each practice shares common characteristics and are likely to impact assets in similar ways.

Validation Authority (VA) An independent organization that provides proof of identity for the requestor who needs to ensure a digital certificate is valid.

Chapter 9 List of Terms

Apache Spark A computational engine used in Hadoop environments.

Apache Storm A real time data processing tool engineered to operate in environments where data streams have no discrete ending point. This is ideally suited for IoT or situations where data may sporadically appear in rapid spates.

Application Services A fast growing area in cloud computing that often falls under the category of SaaS and includes application programs used to accomplish anything that an end-user might wish to do. In general, cloud application services run from servers located in a cloud-based data center. Most often, application services are off-premise and are shared resources used by large numbers of end-users which are permitted use of resources according to their authorization level.

Cloud Analytics Services Use of data analytics tools and services in cloud environments.

Collective Accountability Describes that a team has a common goal: a healthy IT system for the organization.

Compute Services Cloud services that focus on applications requiring CPU intensive operations as opposed to data intensive operations. Compute tasks in the cloud are driven by virtual central processing units (CPUs), arithmetic processing units (APUs), and graphical processing units (GPUs).

Continuous Improvement Continuously finding problems or weaknesses and working to resolve them with long-term, forward-thinking solutions.

Data Analysis Expressions (DAX) A formula language that helps an analyst construct customized expressions to facilitate building "Calculated Columns" and "Measures" in Power BI and other tools. DAX includes many functions like those found in MS-Excel but goes beyond that with specific commands to deal with both relational data and aggregations for many data types. DAX enables the construction of complex statements and filtering logic.

Database-as-a-Service (DBaaS) Cloud vendor runs and maintains the database system on their infrastructure and the cloud user obtains accounts and instances of the database for their organization's use. DBaaS vendors manage updates, problems, backups, and security issues. Cloud users pay for use, often on a per access basis.

DevOps Integration between an organization's software development practices and its IT operations. The cloud enables developers to work closely with those running their software to automate the process of creating and deploying new or updated applications. The key to DevOps is collaboration: software developers and IT teams build, test, and release software in reliable and rapid ways that focus on teamwork and integration.

DMAIC An acronym widely used due to the popularity of Six Sigma, which incorporates this approach to problem-solving. Essentially, it is a roadmap useful for organizing quality improvement processes or entire projects. The letters in the acronym "DMAIC" represent the five primary steps used to solve a problem. These are: Define, Measure, Analyze, Improve, and Control.

Domo Both a software company and product used for business intelligence and visualization. It suggests that data holds stories and the tool enables analysts to extract and communicate those stories. Domo features a central dashboard for managing user experiences.

Flow in DevOps Value-adding work that continually occurs without delay between process steps. Each activity enhances the overall project in the shortest time frame possible.

Hadoop An open source data processing and storage framework for distributed systems. Hadoop often provides the central data processing infrastructure for organizations engaged in data mining, business analytics, predictive analytics, machine learning, and other techniques that involve big data applications.

Hadoop Common One of Hadoop's primary modules that provides a Java-based toolset which analysts use for reading data or other tasks suited to the end-user's computer system.

Hadoop Distributed File System (HDFS) One of Hadoop's primary modules that supports high speed access for application data. Hadoop's objective is to ensure large amounts of data can economically be stored across low-cost distributed computing systems. HDFS stores the data in a manageable form on linked storage devices.

Lean Computing Means "less is more" and cloud resources should only be deployed to the level needed and no more.

MapReduce One of Hadoop's primary modules that performs two basic tasks. One is pulling data from the database and the second is mapping the data into a form that suits the current analysis which could mean performing an operation on the data (e.g. counting or averaging values). So, putting into a form suitable for the analysis is the *map* part and performing an operation *reduces* the data.

Microservices Restructuring application architecture in cloud environments so software solutions are composed of numerous small services. Each service is independent and provides a result or requests information via HTTP or REST API.

Multi-Model Database Relational database systems that also incorporate NoSQL features. This means they added JSON, key-value, graph, document storage capabilities, and other options.

NoSQL Database Database systems that do not use relational tables. Instead, these systems were developed for cloud-based environments and therefore incorporated features to deal with scalability and heavy read/write demands.

OpenStack Overview A popular, open-source cloud software platform (sometimes called a cloud operating system).

Quality Ensuring robustness in cloud computing. Comprises lean approaches, continuous improvement, and agile methods.

SQL Database Database systems that rely on structured tables with rows and columns linked together via relationships composed of unique, shared key fields. Traditional computing applications use relational database structures and SQL commands to retrieve and manage data.

SQL Queries Commands retrieve data and provide it to users or application programs in relational database systems.

Tableau A popular visualization software system widely used in sciences, business, economics, and other areas. Tableau offers visualizations that permit users to drill down into the data with increasing levels of detail.

Theory of Inventive Problem-Solving (TRIZ) A problem solving approach developed in the former Soviet Union. Its overall perspective suggests creativity is at the heart of innovation and should be structured for use in organizations. The structure makes problem solving predictable and helps in group situations. TRIZ posits that most problems have been solved. If not, a similar problem has been solved. The key, then, becomes finding the solution and customizing it to the current problem in a creative manner.

Traditional Database Running in a Cloud Environment Database software is installed and run on cloud infrastructure.

Virtual Machine Database Database installed using a virtual image of the software provided by the vendor. It will probably be optimized for their environment.

Yet Another Resource Navigator (YARN) One of Hadoop's primary modules that supports, manages, and controls the storage resources available to Hadoop.

Chapter 10 List of Terms

Cloud Computing as a Utility The idea that cloud computing will become even more like public utilities as time goes on, both for private individuals and organizations. The model is like electric utilities or city water metering. As computing resources are consumed, a monthly fee is assessed and paid by the user. The shared resources reduce the costs for everyone in the long run.

Cloud Streaming Services that enable and provide consumers with rich media from cloud-based sources. Entertainment videos, music, television channels, virtual reality applications, video games, online courses, digital books, and other material are examples of content that is streamed.

Cloudlet Computing (CC) A leading implementation of Mobile Cloud Computing (MCC) to enable IoT's integration with the cloud. It is essentially a trusted, small cloud infrastructure, located at the edge of the network available to nearby mobile devices.

Edge Computing A layer of computing power that enhances the nexus between cloud computing and IoT. In its most basic form, edge computing relies on technologies that allow computation to occur on a network's edge – usually on downstream cloud data and upstream IoT data. Basically, essential data processing occurs closer to IoT devices or other sources via a mobile, Bluetooth, or wireless network.

Everything as a Service (EaaS) Represents the idea that every computing application can be redeployed and refit to run in the cloud. This mostly relates to application software and consumer products such as video and music streaming and mobile apps.

Fog Computing Takes Edge computing to a higher level by offering distributed computing, storage, control, and networking capabilities closer to the user. It is not limited to the network's edge, instead it provides a structured intermediate layer that fully bridges the gap between IoT and Cloud computing.

Internet of Things (IoT) Enhancing everyday devices and sensors with embedded computing capabilities then collecting data via Internet connections. Devices used range from personal fitness bands to home assistants like Amazon's Echo to GPS devices tracking automobile movement.

Machine Learning Software with an artificial intelligence basis used in data analytics. It becomes more accurate in predicting outcomes without the intervention of humans.

Mobile Cloud Computing (MCC) A cloud processing model that suggests mobile devices offload storage and computation to reduce the mobile device processing load. This idea advocates processing and storage on additional devices at a network's edge. This enables pre-processing prior to data moving into a broader cloud environment.

Mobile Edge Computing (MEC) An approach to Edge computing that brings computation and storage to a mobile network's edge via a Radio Access Network. This offers context awareness, scalability, location responsiveness, low latency, and high bandwidth potential.

Multicloud This approach to cloud computing is a business model where resources from a variety of a variety of cloud vendors are required and used.

NoOps An abbreviation for "No Operations." This refers to the idea that IT infrastructures in organizations could become so automated that operations personal would no longer be needed.

SCALE An acronym used to describe Fog computing benefits regarding the interface between IoT and Cloud Computing. These advantages include Security, Cognition, Agility, Latency, and Efficiency. These represent how preprocessing closer to the user or device improves cloud operations.

Serverless Architecture A computing model consistent with the idea of NoOps, where cloud clients do not need to worry about day-to-day operations in cloud data centers. The main concepts include automating the level of infrastructure needed and then provisioning or deprovisioning resources required to keep operations running optimally.

Small Business Clouds Cloud computing concepts utilized by small and medium enterprises (SMEs) to enhance their business operations.

Supervised Learning A machine learning approach which includes a significant human dimension in the process where a data scientist provides feedback to the model as it develops and determines what variables should influence the learning behavior. Supervised learning involves a training phase prior to use on a full data set.

Unsupervised Learning A machine learning approach that seeks unknown patterns in data sets with no pre-existing categorization nor direct human guidance.

Zero Knowledge Cloud Storage All data is encrypted prior to movement into the cloud so that the cloud provider and anyone working in their data center cannot access the data in a meaningful or useful way.

Index

Cloud Technologies: An Overview of Cloud Computing Technologies for Managers, First Edition. Roger McHaney.
© 2021 John Wiley & Sons Ltd. Published 2021 by John Wiley & Sons Ltd.
Companion website: www.wiley.com/go/mchaney/cloudtechnologies